纺织服装类"十四五"部委级规划教材

Clothing
and Its Production

服装英语

（第3版）

卓乃坚　　[英] 西蒙 C. 哈罗克　　著
Zhuo Naijian　　Simon C. Harlock

东华大学 出版社 · 上海

图书在版编目(CIP)数据

服装英语 / 卓乃坚,(英)西蒙 C. 哈罗克著. —3 版.
—上海:东华大学出版社,2022.8
ISBN 978-7-5669-2090-4

Ⅰ.①服… Ⅱ.①卓… ②西… Ⅲ.①服装工业—英语
Ⅳ.①TS941

中国版本图书馆 CIP 数据核字(2022)第 134043 号

责任编辑:谭 英
封面设计:张林楠 Marquis

服装英语(第3版)

Fuzhuang Yingyu Disan Ban

卓乃坚 [英]西蒙 C. 哈罗克 著

东华大学出版社出版
地址:上海市延安西路 1882 号
邮政编码:200051 电话:(021)62373056
出版社官网 http://dhupress.dhu.edu.cn/
出版社邮箱 dhupress@dhu.edu.cn
上海龙腾印务有限公司印刷
开本:787mm×1092mm 1/16
印张:14.25 字数:431 千字
2022 年 8 月第 3 版 2022 年 8 月第 1 次印刷
ISBN 978-7-5669-2090-4
定价:45.00 元

内容提要

本书用英语介绍了服装的分类、衣片和样板、服装的面料和辅料、服装生产工艺流程和设备以及服装的质量问题等。本书在强化专业术语应用的同时，注重了与服装出口贸易的关联，并配有练习、词汇检索以及中文参考译文等。每章的末尾载有精心挑选出来的摘自国外有关杂志或网站的阅读材料，配上必要的参考提示，可供读者在学习课文内容后，进一步扩大自己的知识面。本书可作为普通高等院校服装专业的专业英语教材或国际经济与贸易专业的服装外贸之类选修课程的双语教材，也可作为服装外贸从业者一本不可多得的参考读物。

Preface

There is considerable uncertainty in today's world. In the recent years, the incertitude seems to grow further due to the impact of COVID-19 epidemic. However, we should say that one thing is still for sure, that is, English is still one of the important tools for people making communication around the world. Therefore, for people being involved in the field of clothing making or garment trading it would be very beneficial to grasp the specialized English terminologies regarding the clothing field, if they wish to know further about the clothing world and hope to push their products towards the world market.

In this present edition, we have updated all reading materials in the hope that after getting the fundamental knowledge from the texts in this book our readers could get more fresh knowledge about how people worldwide in clothing field use English. Our purpose of learning English was to use English. Necessary reference tips are given to those elaborately selected reading materials to facilitate our readers to understand those materials.

Finally, special thanks are given to Mr. Zhuo Shufan for all work he has done in retrieving and preparing the reading materials adopted in this edition.

The authors
February 2022

前言

当今世界充满变数,近年来的新冠疫情似乎使得一切变得更不确定。然而,我们应该说,有一件事仍然非常肯定,即英语仍然是人们在国际交往中的重要工具。因此,对于从事服装生产或贸易的人来说,如果想要更好地了解服装世界,把自己的产品推向世界,掌握服装方面的专业英语术语将会非常有益。

本书这一版更新了所有阅读材料,以便读者能够在学习了本书课文介绍的基本知识后,对其他国家的人们在服装领域如何使用英语有更新的了解。我们学习英语的目的,就是为了使用。为了方便读者理解,所有精心挑选的阅读材料都加上了必要的参考提示。

最后,特别感谢卓书帆为检索和准备本版阅读材料所做的所有工作。

作者
2022 年 2 月

Contents

目　录

CHAPTER 1　CLOTHING CLASSIFICATIONS　服装分类

1　Classifications with a View to Everyday Life 日常分类　　1

1.1　Upper Clothing 上装　　1

1.2　Lower Clothing 下装　　3

1.3　Suits and Ensembles 西服套装和便服套装　　4

2　Classifications with a View to International Trade 国际贸易中的分类　　5

2.1　Classification Based on H. S. Code 基于协调制编码的分类　　5

2.2　Classification Based on Special Regulations 基于特殊规则的分类　　6

2.3　Classification Based on the Method of Manufacture 基于生产方法的分类　　6

3　Importance of the Classifications 分类的重要性　　6

3.1　For Statistical Purposes 为了统计　　7

3.2　To Describe the Name of the Goods in a Contract 为了描述合同中的商品名称　　7

Words and Phrases 词汇　　11

Exercises 练习　　17

Reading Materials 阅读材料　　18

CHAPTER 2　GARMENT SECTIONS, PATTERNS AND MEASUREMENT LOCATIONS　服装组成部分、样板和测量部位

1　Garment Sections 服装组成部分　　21

1.1　Sections in Upper Garments 上装的组成部分　　21

1.1.1　Body 大身　　21

1.1.2　Sleeves 袖　　23

1.1.3　Collars 领子　　24

1.1.4　Pockets 衣袋　　25

1.2　Sections in Lower Garments 下装的组成部分　　26

1.2.1　Trousers 裤　　26

1.2.2　Skirt 裙　　27

1.3　Other Garment Components and Styling Details 其他衣片和款式细节　　28

2   **Patterns** 样板   30

3   **Measurement Locations** 测量部位   31

Words and Phrases 词汇   34

Exercises 练习   41

Reading Materials 阅读材料   42

CHAPTER 3   GARMENT FABRICS, ACCESSORIES, LABELS AND HANGTAGS   服装的织物、辅料、标签和吊牌

1   **Garment Fabrics** 服装的织物   45

   1.1   Shell Fabrics 面料   45

   1.2   Linings 里料   46

2   **Accessories** 辅料   47

   2.1   Sewing Threads 缝纫线   47

   2.2   Fasteners 扣合件   49

   2.2.1   Buttons 纽扣   49

   2.2.2   Zippers 拉链   49

   2.2.3   Snaps 揿纽   51

   2.2.4   Velcro 尼龙搭扣   51

   2.3   Interlinings 衬头   52

   2.4   Wadding 填料   52

   2.5   Other Accessories 其他辅料   53

3   **Labels and Hangtags** 标签和吊牌   54

   3.1   Labels 标签   54

   3.1.1   Brand Label 品牌标签   54

   3.1.2   Origin Label 产地标签   54

   3.1.3   Size Label 尺码标签   55

   3.1.4   Washing Label 洗涤标签   55

   3.1.5   Other Labels 其他标签   55

   3.2   Hangtags 吊牌   56

Words and Phrases 词汇   57

Exercises 练习   63

Reading Materials 阅读材料   64

CHAPTER 4   WORKING SHEETS   工艺单

1   **The Function of Working Sheets** 工艺单的作用   68

2   **Contents of a Working Sheet** 工艺单的内容   68

   2.1   Material Specifications and Consumption 材料的规格和耗用量   69

   2.2   Working Sketch 效果图   70

2.3   Technical Details 技术细节   70

2.3.1   Stitch Types 线迹类型   70

2.3.2   Stitch Density 线迹密度   74

2.3.3   Seam Types 缝子类型   75

2.3.4   Edge Binding and Seam Neatening 滚边和缝子毛边处理   76

2.3.5   Seam Allowance 缝头   77

2.4   Size Chart 尺寸表   77

3   **Abbreviations and Simplified Words in the Working Sheets** 工艺单中的缩略词
和简化词   78

Words and Phrases 词汇   82

Exercises 练习   84

Reading Materials 阅读材料   86

CHAPTER 5   GARMENT MANUFACTURE   服装生产

1   **Introduction to Manufacturing Sequences** 生产工艺流程介绍   90

1.1   Manufacturing Sequences for Cut-fashioned Garments 裁剪成形服装的
生产流程   90

1.1.1   Pattern Cutting 样板制作   90

1.1.2   Sample Making 样衣制作   92

1.1.3   Grading 推档   92

1.1.4   Fabric and Accessory Inspection 织物和辅料检验   92

1.1.5   Lay-planning and Marker-making 铺料设计和画排料图   94

1.1.6   Fabric-spreading 铺料   97

1.1.7   Cutting 裁剪   98

1.1.8   Sorting, Bundling and Numbering 分类、打捆和打号   98

1.1.9   Fusing 上黏合衬   98

1.1.10   Sewing, Under Pressing and Trimming 缝纫、小烫和修整   99

1.1.11   Finishing and Final Pressing 整理和成衣熨烫   99

1.1.12   Garment Inspection 服装检验   99

1.1.13   Packaging 包装   100

1.2   For Knit-fashioned Garments 对于编织成形的服装   100

2   **Equipment in Clothing Factories** 服装厂设备   102

2.1   Equipment in Cutting Rooms 裁剪车间设备   102

2.1.1   Cutting Table 裁剪桌   102

2.1.2   Spreading Machine 铺料机   103

2.1.3   Cutting Tools 裁剪工具   103

2.1.4   Fusing Equipment 上黏合衬设备   104

2.1.5　Other Equipment in Cutting Rooms 裁剪车间的其他设备　　　105

2.2　Equipment in Sewing Rooms 缝工车间设备　　　105

2.2.1　Overview of Primary Sewing Machines 基本缝纫机的概况　　　106

2.2.2　Primary Systems in a Sewing Machine 缝纫机的主要系统　　　106

2.2.3　Other Equipment 其他设备　　　109

2.2.4　Production Lines 生产线　　　109

2.3　Equipment Used in the Finishing Room 整理车间所用设备　　　111

Words and Phrases 词汇　　　112

Exercises 练习　　　120

Reading Materials 阅读材料　　　122

CHAPTER 6　GARMENT QUALITY ISSUES　服装的质量问题

1　**The Importance of Garment Quality** 服装质量的重要性　　　125

2　**The Criteria for Garment Quality** 服装的质量准则　　　125

3　**Common Garment Quality Problems** 常见的服装质量问题　　　126

3.1　Inherent Quality 内在质量　　　126

3.2　Apparent Quality 外在质量　　　128

3.2.1　Apparent Quality Problems due to Fabric Faults 源于织物疵点的外在质量问题　　　128

3.2.2　Apparent Quality Problems due to Poor Workmanship or Manufacturing Skills 源于不良的做工或生产技术的外在质量问题　　　129

Words and Phrases 词汇　　　131

Exercises 练习　　　133

Reading Materials 阅读材料　　　134

Keys to Exercises 练习题答案　　　137

Chinese Version 中文译文　　　138

Index to Words and Phrases 词汇索引　　　191

CHAPTER *1* CLOTHING CLASSIFICATIONS

1 Classifications with a View to Everyday Life

" Clothing" may be classified in several ways. For example, everyday clothing may be classified as underwear and outerwear according to the layer in which the clothing is worn. Underwear, or underclothing, is the clothing worn close to the body, i. e. under other clothes, and outerwear, as the name implies, refers to clothes worn over underwear. However, nowadays it may be difficult to define whether a piece of clothing is to be considered as underwear or outerwear. For example, a singlet top could be worn by a girl as either underwear or "outerwear" during the summer season, and a T-shirt, which was historically considered to be underwear, is now widely accepted as an outwear worn in public.

Another way commonly used to classify clothing is to classify an item of clothing as either a "top" garment (upper garment) or a "bottom" garment (lower garment) according to the "half" of the body on which it is worn, i. e. the "top" half from the neck downwards notionally to the waist or the "bottom" half downwards from the waist to the feet. Attention should be paid to the fact that, within a single garment that covers most of the body, the upper part of the garment can be referred to as the "garment top" instead of a "top garment" and the lower garment can be referred to as the "garment bottom" instead of "bottom garment". The terms "top garment" and "bottom garment" can lead to further confusion because "top garment" can also mean an outer garment and "bottom garment" an under garment. Furthermore, the term "top" or "bottom" can simply be used on its own to refer to the upper garment or lower garment.

1.1 Upper Clothing

If one observes a clothing ensemble from the outside to inside or from top to bottom, it could be seen that a commonly worn outerwear top is a coat. The word "coat" is actually a very general term used to name a garment usually with long sleeves and with a length anywhere from just above the knee to the ankle. According to their styles, the fabrics involved and their usage, etc., there are over coats, top coats, duster coats, trench coats, casual coats, rain coats and all-weather coats, etc. Sometimes, the word "coat" is not necessarily used to describe such kind of garments. For example, the "parka" is a kind of casual coat.

The jacket is also a kind of outerwear top that may be worn under a coat. Broadly speaking, the "jacket" is the general name for almost all short upper outer garments. A jacket can be an item of either formal or casual wear. In formal wear, a jacket is a tailored garment, single or double breasted, with or without vents at the back, which, if included, may be either single or double vented and with lapels. These may be worn as part of a suit, usually with matching trousers, or as a separate item of clothing worn with either more formal or casual trousers. A formal dinner jacket is one worn on formal occasions, whereas a lounge or "everyday" suit is one that would be worn on less formal occasions or for office work. The blazer is a jacket with a lapel, usually double breasted, which, although traditionally associated with a uniform worn, for example, by airline flight crew, school children and members of yacht clubs, is increasingly worn by the general public as an outerwear jacket. A casual "jacket" in its narrowest sense is a long sleeved casual short outerwear top usually with waistband or with a drawstring waist. The blouson is a kind of loose-fitting jacket usually with large armholes and elasticated waist. A garment in a "jacket" style with a relatively longer garment length would be classified as a casual coat in the trade. Garments, such as wind-cheaters, anoraks and snorkel parkas, usually fall under the coat and jacket category. Therefore, the garment length, one of the main factors, determines which category the garment should fall into.

Sweaters are knitted garments, which can usually be worn alone or under a jacket or coat. Disregarding the designs or loop/stitch structures, a sweater may be classified as a pullover (because it is worn by pulling it on over the head) and a cardigan (the same word is used for a type of knitted structure with tuck stitched courses), which is buttoned down the front. The pullover is also called a "jumper" in British English. In contrast, in American English, a "jumper" (or "jumper dress", which in turn in British English is called as "pinafore dress") could mean a sleeveless dress usually worn over a blouse.

A shirt, which is worn by both men and women, may be an item of formal office clothing, usually worn together with a tie. In America, such a shirt would be called a "dress shirt". Traditionally, formal shirts are made of woven fabrics, with collar, yoke and cuffs. Typically, a shirt collar is composed of collar and collar stand with interlining applied to offer a stiff collar shape, but nowadays the shirt with a band collar can be found everywhere, especially in some commercial offices. Compared to the shirt, the blouse is usually only worn by women (but in America, the word "blouse" refers also to a style of men's military uniforms). Blouses may be either long or short sleeved with a flat collar or tie collar, or even without collar and, sometimes, sleeves. Lace may feature on the collars or cuffs to impart more femininity to the garment. Blouses may have many sleeve or collar styling variations, but for shirts, the style of the shirt, especially the sleeves, is more or less stereotyped into short or long with a sleeve vent and buttoned cuff. It should be pointed out that a garment name containing the word "shirt" does not always refer to the style of shirt mentioned above, and the T-shirt and the sweatshirt might be the good examples. These are usually knitted with a round neck and without collar and pockets. The sweatshirt is usually produced from knitted fleece fabric with long sleeves, but the T-shirt usually has short sleeves. Nowadays, the zipper hooded sweatshirt is very popular among young people. A polo shirt is a casual, usually cotton, knitted short sleeve shirt with a buttoned front opening, large enough to allow it to be pulled on over the head. A sleeveless T-shirt could be called a tank top.

The word "dress" can be a countable noun or an uncountable noun. As a countable noun, the word "dress" applies only to women's garments. "Dress" is a generalized word, and any skirt with an attached top as a single piece could be called a "dress". There are many varieties of dresses, such as the pinafore dress (which, as mentioned earlier, is called as "jumper dress" in

American English) and the sundress. As an uncountable noun, the word "dress" sometimes forms a compound expression with other words to mean one special kind of clothing, such as national dress, evening dress and working dress, and under such circumstances, such expressions could be used for men's or women's garments. To "dress" oneself is to put clothes on.

The most common underwear top is the vest or singlet. By definition, a vest is usually an upper garment without sleeves. However it does not necessarily refer to underwear; for example, a waistcoat is referred to as a vest (in America). According to its general meaning, a vest is a kind of underwear top; therefore, beside the sleeveless vest, there could be long-sleeved vests or short-sleeved vests.

For women, a "slip" is a type of underwear top (and, of course, there could be half slips). In addition, there are foundation garments, such as brassieres (bras), corsets and camisoles; all are classified under the general category of lingerie.

1.2　Lower Clothing

One of the most common lower outer garments is trousers, and in American English, the term pants or slacks instead of trousers is used. However, the term pants in British English refers to lower underwear (presumably abbreviated from underpants i. e. worn under pants), which, in turn, would be called underpants in American English.

The term trousers is an all-embracing term, and the distinction is that formal trousers are part of a suit or worn to the office, and workwear trousers are worn in more manual working environments or casual trousers for leisure wear. Some terms nowadays have emerged such as "chinos" for a style of casual trousers and, of course, jeans. The term slacks is the one used usually to describe ladies and (in the USA) men's casual trousers which, often but not always, have elasicated waists for comfort.

Trousers usually extend from the waist to the feet. "Trousers" that end above or below the knee are called shorts. Shorts worn for swimming are called trunks, i. e. swimming trunks. Breeches, jeans and "bib and brace" overalls, etc. are common styles of trousers. The silhouette may be changed to produce different styles of trousers, of which trousers with tapered, flared or straight legs are typical examples. Usually Bermuda shorts and some trousers have "turn-ups" on the leg hems. Other changes in pockets or waist etc. can also create new styles of trousers. Slacks and trousers that have ankle vents with zipped closures are very fashionable casual outerwear bottoms nowadays. If used as sportswear, they may be referred to as "jogging" bottoms. Stonewashed jeans with whisker hand brushing are also very popular with young people.

Skirts are lower garments for women. However the Scottish kilt worn by men is a kind of "skirt", traditionally worn by Scottish male soldiers, which nowadays seems to be worn only on some traditional ceremonial occasions or at sports venues to emphasize patriotism for Scotland. Furthermore, the saree and the sarong are also kinds of wrapped-and-tied or tucked-in skirts worn commonly by indigenous women and some tourists in tropical countries. Likewise, men also wear a similar style of garment in such countries, although they tend to be referred to by a national name rather than sarong. There are also various styles of skirts, such as A-line skirt, straight skirt, peg top skirt, pleated skirt, gored skirt, mini-skirt and wrap-round skirt. Clearly, by changing its length or silhouette, or by adding some styling details, considerable varieties of skirts can be designed. Other examples of skirt styling are skirts with a yoke, skirts with a high waist or a dropped waist, and skirts without a waistband, etc. Wider legged shorts are referred to as culottes or divided skirts.

As mentioned previously, pants or underpants are widely used words for underwear "bottoms". The word "briefs" is a very common term used to refer to underwear bottoms for both male and female. However, in addition to briefs, more commonly-used terms for girls and ladies are pants, panties or knickers. Though, technically speaking, only longer briefs would be called knickers (British English, or panties in American English), the term "knickers" is used interchangeably with "pants" in the U. K. It should be noted that the word "knickers" can also be used as an abbreviation in the USA for knickerbockers, somewhat old-fashioned baggy knee trousers usually for men or boys. Thongs and bikini bottoms refer to revealing female underwear, the latter derived from the two-piece bikini swimsuit. Underwear for men is defined by styling features such as Y-fronts and boxer shorts. Tights or panty-hose are worn only by women.

1.3　Suits and Ensembles

Upper and lower garments are invariably sold separately. Women usually like to buy separates, and with carefully chosen combinations of the upper and lower garments, they can create various fashionable styles that often reflect their personalities.

A set of garments composed of two or three pieces and made up from, usually, identical fabric with the same styling features would be called a "suit". A suit usually consists of a jacket and a pair of trousers, or for women, a jacket and a skirt. A three-piece suit includes a waistcoat (which, as mentioned before, is also called a vest in America). Generally, the two-piece or three-piece men's suit or women's suit is usually worn either on formal occasions or as office-wear, and, as the traditional western standard business wear for men, the suit is usually worn with a long-sleeved shirt and a tie. Even more formal wear is the dinner suit or dinner jacket, worn on very formal occasions, or the morning/wedding suit sometimes worn at ceremonial receptions or by bridegrooms at weddings. However, there are other "suits", such as a track suit or knitted suit, which are more casual wear. If a garment (not a dress) is designed with the top and bottom in one piece, the word "suit" would also be used to form a compound word to describe the garment such as "jumpsuit" and "body suit". "Ensemble" or "set", similar words other than "suit", would be used to describe a set of garments with top and bottom components coordinated in some way.

In addition to the above classifications, there are some other ways to classify or name a garment. Some garment names are derived from their original usages, for example, sportswear, tennis shirt, duster, beachwear and evening gown. Some garment names come from the ones of their original designers or follow the name of the place where they are popular, such as the rain coat referred to as a Mackintosh (The inventor's name was Charles Macintosh), Arabic robe, and Hawaiian shirt. In international trade, it is usually necessary to specify whether the garments are for men, women or children, for example, men's coats (or gents' coats), women's coats (or ladies' coats) or children's coats. How a name is given to a garment is more or less due to tradition and people's preference.

2 Classifications with a View to International Trade

Garment trade is one of the important fields in international trade, especially between the developed and developing countries. In order to monitor the garment trade, people have established formally recognized classification systems for garments. This enables statistics regarding the import and export of garments to be compiled by the Customs and Excise Offices in the respective countries.

2.1 Classification Based on H. S. Code

One of the best ways to classify garments is by using the Harmonized Commodity Description and Coding System (i. e. Harmonized System), abbreviated as the H. S. Code.

In the Harmonized Commodity Description and Coding System, textiles and textile articles fall into Section XI. Knitted or crocheted garments are generally classified within Chapter 61 and garments that are not knitted or crocheted, or not made of knitted fabrics, are in Chapter 62. Some other chapters also relate to garments, and for example, second-hand or worn clothing falls into Chapter 63, and fur clothing and the garments with fur linings fall into Chapter 43.

Attention must firstly be paid to the notes in every chapter, which give specific explanations to the application of the chapter. For example, the word "suit" is strictly defined in the notes for Chapters 61 and 62. According to these notes, the term "suit" means a set of garments composed of two or three pieces made up, in respect of their outer surface, in identical fabric and comprising:

☐ "one suit coat or jacket, the outer shell of which, exclusive of sleeves, consists of four or more panels, designed to cover the upper part of the body, possibly with a tailored waistcoat in addition, whose front is made from the same fabric as the outer surface of the other components of the set and whose back is made from the same fabric as the lining of the suit coat or jacket; and, "

☐ one garment designed to cover the lower part of the body and consisting of trousers, breeches or shorts (other than swimwear), a skirt or a divided skirt, having neither braces nor bibs. "

Furthermore, the notes stress that all of the components of a "suit" must be of the same fabric construction, colour and composition; they must also be of the same style and of corresponding or compatible size. However, these components may have piping in a different fabric.

According to these notes, the term "suit" also includes the following sets of garments whether or not they satisfy all the above conditions:

☐ morning dress, comprising a plain jacket (cutaway) with rounded tails hanging well down at the back and striped trousers;

☐ evening dress (tailcoat), generally made of black fabric, the jacket of which is relatively short at the front, does not close and has narrow "tails" cut in at the hips and hanging down

5

behind;

☐ dinner jacket suits, in which the jacket is similar in style to an ordinary jacket (though perhaps revealing more of the shirt front), but has shiny silk or imitation silk lapels.

Some other terms, such as "ensemble" and "set", have also been strictly defined in the notes of Chapters 61 and 62.

2.2 Classification Based on Special Regulations

In order to monitor the garment importation some countries or regions set up their own systems to classify garments. There can be big differences between these systems. In the E. U. system and in the Canadian system, garments are classified according to both the type of garments and the type of constituent fabric, for example, trousers of woven fabric are cited under Category 6 in the E. U. and under Category 5 in Canada.

In the United States, the category system is quite different and a three-figure category number is used. The first figure denotes the textile material used in the garment (for example, 2 for garments made of cotton and/or man-made fibres; 3 for garments made of cotton; 4 for wool garments; 6 for garments made of man-made fibres and 8 for garments made of silk blends or non-cotton vegetable fibres), and the other two figures denote the type of garment. For example, cotton men's trousers come under Category 347 and women's cotton trousers come under Category 348. If the same trousers are made of wool, they will fall into Category Nos. 447 and 448.

2.3 Classification Based on the Method of Manufacture

In international trade, it is very important to identify the country of origin, because the Customs and Excise Office of the importing country decides what kind of duties should be levied according to the country of origin and the government trade policies. It is simple to determine the country of origin if all the manufacturing processes of a garment occur in one country and all materials used are local, but if imported materials are involved or the garments are made up in different countries, the regulations on how to identify the country of origin become important.

Countries implement different regulations to determine the country of origin for garments, but most countries set up their regulations to determine the origin according to where the "main process" in their production happens. However, what the "main process" is seems to be a controversial issue. Many countries would classify garments as cut-fashioned or knit-fashioned. For the cut-fashioned garments, the country where the cut garment sections are assembled is defined as the country of origin, and for the knit-fashioned garments, the country where the garments or their fashioned pieces (through widening or narrowing) are knitted is defined as the country of origin. It must be noted that it is quite possible for a country to adjust its criterion for origin according to the current situation of garment importation. To standardise the regulations on the country of origin issue is a great expectation for the World Trade Organization.

3 Importance of the Classifications

The method of classification of garments is less important in everyday life; however in business transactions, especially in the international trade, it is essential to correctly classify and name the garments for the following reasons.

3.1 For Statistical Purposes

The relevant authorities in every country have to compile statistics for the imports and exports, so that, based on the statistical data, the government may adjust its import and export policies. However, without scientific ways to classify the garments, statistics for garment trade could not be generated efficiently. No matter whether it is the Harmonized Coding System, the E. U. category system, or the U. S. category system, their intention in providing a recognized classification system is the same.

The governmental or regional regulations regarding the country of origin will affect the statistics. For example, according to the Harmonized Tariff Schedule of the United States, garments manufactured (cut and assembled) from fabric formed in the United States may be eligible for entry under the Outward Processing Program for textiles and apparel. However, the eligibility must be in compliance with procedures established by the Committee for the Implementation of Textile Agreements (CITA). The importer is required to identify such garments on the entry summary forms by placing the symbol "S" as a prefix to the appropriate 10-digit harmonized tariff number.

3.2 To Describe the Name of the Goods in a Contract

The name of a garment is part of the description of goods in a garment import or export contract. Therefore, if people know how the garments are classified, they would know how to name the garments for such contracts. However, attention should be paid to the garment name before it is settled or agreed in a contract.

Firstly, either the same or a similar garment name might have a different meaning in different regions or countries. As mentioned previously, the word "pants" refers to a lower outer garment in American English, but usually refers to a lower under garment in British English. The so-called "T-shirt" in China usually refers to a kind of knitted shirt with a collar instead of the more common sense definition, that is, a short-sleeved collarless round neck knitwear which, when opened flat, resembles the letter "T". A knitted shirt or knitwear with a collar should fall within the classification corresponding to H. S. 6105 or H. S. 6106 whereas the T-shirt, which is knitwear without a collar, should fall into H. S. 6109 in the harmonized coding system.

Secondly, in international trade, English is the widely used language for business documentation. Therefore, it is very important to use the words commonly used by the English native speakers to describe the garment in the contract. Any new expressions derived just from the imagination of a person whose mother tongue is not English would possibly cause confusion, and potentially lead to a subsequent trade dispute. Attention to this point should also be paid when sellers send price lists to their potential customers or, buyers send their enquiries for quotations.

Thirdly, if the materials involved are included in the garment descriptions, the correct components in the materials should be given. However, in the harmonized coding system, fabrics of a mixture of two or more textile materials should be classified as if consisting wholly of the textile material that predominates by weight rather than each single textile material. For example, a pullover knitted from a blend of 85% wool and 15% Acrylic yarns will be classified under H. S. 6110. 11 as a pullover made of wool, but in a contract the correct blend ratio must be given. It is never advisable to simply describe the garment as a "Wool Pullover" instead of the more precise description of a "Wool/Acrylic 85/15 Pullover" in a contract.

Fig. 1. 1　Coats
1 — double breasted overcoat　2 — duster coat
3 — trench coat　4 — top coat　5 — parka

Finally, if possible, the garment name in an import or export contract should be in line with the corresponding name used in the harmonized coding system, or with the name under the corresponding agreement category if the trade is restricted by some bilateral governmental agreement.

If multiple styles of garments are involved in a transaction, the contract could name the garments in general terms and then specify the quality requirements for the shell, lining and accessories and list the style numbers with the corresponding quantities and the unit prices. Of course, no matter how the garments are to be named in a contract, samples or working specification sheets (SPEC sheets) may be used to clarify exactly the products involved. Mutual understanding is essential for the smooth execution of a contract.

Fig. 1. 2　Jackets
1 — blouson　2 — blazer　3 — anorak　4 — wind-cheater　5 — snorkel parka

Fig. 1. 3　Garments that are associated with the word "shirt"

1 － sweatshirt and shorts　2 － T-shirt and bib-brace overalls　3 － overshirt and jeans

4 － jumpsuit　5 － T-shirt　6 and 7 － shirt（dress shirt）

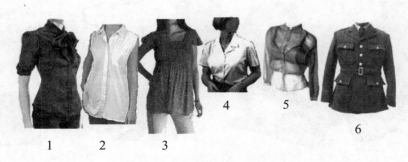

Fig. 1. 4　Blouses

1 － blouse with tie collar　2 － sleeveless blouse　3 － blouse with square neck

4 － short-sleeved blouse　5 － sheer blouse　6 － army dress blouse

Fig. 1. 5　Dresses and skirts

1 － pinafore dress（jumper dress）　2 － sundress　3 － Hawaiian dress

4 － sleeveless dress　5 － jumper and sunray-pleated skirt　6 － skirt suit

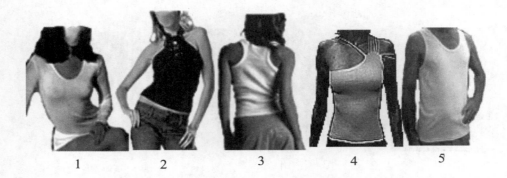

Fig. 1. 6   Vests and singlet tops

1 — long-sleeved vest   2 — halter top   3 — singlet with racer back

4 — spaghetti singlet top   5 — men's singlet vest

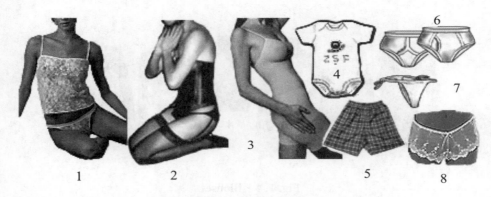

Fig. 1. 7   Underwear

1 — camisole and briefs   2 — corset, panties and stockings   3 and 4 — body suits

5 — boxer shorts   6 — Y-fronts   7 — throngs   8 — knickers

Fig. 1. 8   Skirts

1 — A-line skirt   2 — straight skirt   3 — gored skirt

4 — pleated skirt   5 — wrap-around tennis skirt

Fig. 1. 9　Evening gown, suits, etc.

1－evening gown　2－track suit　3－waistcoat　4－tail coat　5－dinner suit

Fig. 1. 10　Sarong, saree, etc.

1－saree　2－sarong　3－beachwear　4－beach shorts　5－swim suit

Words and Phrases

everyday clothing	日常衣着
underwear [ˈʌndəweə]	内衣
outerwear [ˈaʊtəweə(r)]	外衣
underclothing	内衣
singlet [ˈsɪŋglɪt] top	(女)吊带衫
T-shirt	短袖针织圆领衫
upper garment	上装
lower garment	下装
garment top	上装
garment bottom	下装
top	上装

bottom	下装
coat	风衣，大衣
overcoat	(厚)大衣
top coat	(薄花呢)大衣
duster ['dʌstə] coat	防尘大衣，风衣
trench [trentʃ] coat	军大衣式雨衣
casual ['kæʒuəl] coat	休闲大衣
rain coat	雨衣
all-weather coat	防刮风下雨及日晒的大衣
parka ['pɑːkə]	派克大衣，长夹克式休闲大衣
jacket ['dʒækɪt]	夹克衫，短上装外衣的统称
casual wear [weə(r)]	休闲服
formal wear	正式衣装
tailored ['teɪləd] garment	度身裁制的服装
single breasted ['brestɪd]	单排纽
double breasted	双排纽
vent	开衩
lapel [lə'pel]	驳领
suit [sjuːt]	套装，西服套装
trousers	长裤，裤子
dinner jacket	(没有燕尾的)晚礼服(上装)
lounge [laʊndʒ] suit	普通西服套装
blazer ['bleɪzə]	西便装(式制服)
airline flight crew [kruː]	航空乘务员
yacht [jɒt] club	游艇俱乐部
waistband	腰头
drawstring ['drɔːstrɪŋ]	(用于收紧的)绳带
blouson ['bluːsɒn]	宽松式带橡筋腰的夹克
loose-fitting	宽松的
armhole ['ɑːmhəʊl]	袖窿
style	款式;时尚;风格
wind-cheater	防风上衣
anorak ['ɑːnərɑːk]	带风帽的短风衣
snorkel ['snɔːkl] parkas	(风帽开口较小的)派克衫
category ['kætɪgərɪ]	类别
sweater ['swetə]	(粗)针织衫，针织毛衣;绒衫
knitted garment	针织服装
design	(图案)花型，设计
loop/stitch structure	线圈结构

pullover ['pʊləʊvə(r)]	套衫
cardigan ['kɑːdɪɡən]	开衫
tuck stitched course	集圈横列
jumper	(英)(粗)针织套衫
jumper dress	(美)无袖套头连衣裙
pinafore ['pɪnəfɔː] dress	(英)无袖套头连衣裙
shirt [ʃɜːt]	衬衫
tie	领带
dress shirt	正装衬衫
woven ['wəʊvən] fabric ['fæbrɪk]	机织织物
collar ['kɒlə]	衣领
yoke [jəʊk]	覆肩
cuff ['kʌf]	袖口(袖克夫)
collar stand	领脚(领座)
band collar	立领
blouse [blaʊz]	女衬衣,(美)军上装
long or short sleeved	长袖或短袖的
flat collar	坦领
tie collar	领结领
lace	花边
femininity [femɪ'nɪnɪtɪ]	女性气质
stereotyped ['sterɪətaɪpt]	固定套路的,老一套的
sweatshirt ['swetˌʃɜːt]	长袖运动衫
round neck	圆领
pocket	衣袋
knitted fleece fabric	针织起绒布
polo ['pəʊləʊ] shirt	套头式针织运动衫
sleeveless	无袖的
tank top	无袖圆领衫
dress	连衣裙
sundress ['sʌnˌdres]	(露肩)背带式连衣裙
national dress	民族服装
evening dress	晚礼服
working dress	工作服
vest	背心,内衣上装,西装马夹(美)
singlet	背心
waistcoat ['weɪstkəʊt]	西装马夹
long-sleeved vest	长袖内衣
short-sleeved vest	短袖内衣

slip	衬裙
half slip	下装衬裙
foundation garment	(女)贴身内衣
bra [brɑː]	文胸,胸罩
brassiere [ˈbræsɪə]	文胸,胸罩
corset [ˈkɔːsɪt]	(女)束腰塑身内衣
camisole [ˈkæmɪsəʊl]	(女)胸衣
lingerie [ˈlænʒəriː]	(女)内衣
pants	(美)外衣长裤,(英)长内裤
slacks	宽松裤
underpants	(长)内裤
leisure [ˈleʒə; ˈliːʒə] wear	休闲服
chinos [ˈtʃɪnəʊz]	咔叽休闲裤
jeans [dʒiːnz]	牛仔裤
shorts	短裤
swimming trunks [trʌŋks]	(男)泳裤
bib [bɪb]	护胸
brace [breɪs]	背带
overalls [ˈəʊvərɔːlz]	工装裤
silhouette [ˌsɪlu(ː)ˈet]	轮廓,侧影,造型
tapered [ˈteɪpəd] legs	小裤脚
flared [fleəd] legs	喇叭裤脚
straight legs	直筒裤脚
Bermuda [bə(ː)ˈmjuːdə] shorts	百慕大短裤
turn-ups	翻边
zipped closure [ˈkləʊʒə]	用拉链的闭口
"jogging" [ˈdʒɒgɪŋ] bottoms	慢跑时穿的运动裤
stonewashed	经石洗的
whisker [ˈhwɪskə] hand brushing	须状手擦(俗称"猫须")
skirt	裙
kilt [kɪlt]	苏格兰格子呢裙
saree [ˈsɑːrɪ]	纱丽
sarong [səˈrɒŋ]	纱笼
A-line skirt	A字裙
straight skirt	直筒裙
peg [peg] top skirt	上宽下窄的裙子
pleated skirt	褶裥裙
gored [gɔːd] skirt	拼片裙
mini [ˈmɪnɪ] skirt	超短裙

wrap-round skirt	左右裙前片叠合的裙子，围裹裙
yoke [jəʊk]	裙(或裤)的拼腰
high waist	高腰
dropped waist	低腰
culottes [kju(ː)ˈlɒts]	裙裤
divided [dɪˈvaɪdɪd] skirt	裙裤
briefs [briːfs]	短内裤，三角裤
panties [ˈpæntɪz]	(美)(女)短内裤
knickers [ˈnɪkəz]	(英)(女)短内裤
knickers/knickerbockers [ˈnɪkəbɒkəz]	灯笼裤
thongs [θɒŋz]	(女)布带裤
bikini [bɪˈkiːnɪ] bottoms	(女)比基尼短裤
Y-fronts	(男)(前开裆)短内裤
boxer shorts	平脚裤
tights [taɪts]	紧身裤
panty-hose	连裤袜
separates	上下装分开销售的服装
personality [ˌpɜːsəˈnælɪtɪ]	个性
office-wear	办公室着装
dinner suit	正餐礼服套装
morning/wedding suit	常礼服套装/婚礼服套装
track [træk] suit	运动套装
knitted suit	针织套装
jumpsuit	伞兵装，连裤外衣
body suit	(紧身)连短裤内衣
ensemble [ɑːnˈsɑːmbl]	便服套装
set	套
sportswear	运动服
tennis shirt	网球衫
duster	防尘服
beachwear	沙滩装
evening gown	(裙摆及地的女装)晚礼服
Mackintosh [ˈmækɪntɒʃ]	胶布雨衣
Arabic [ˈærəbɪk] robe	阿拉伯长袍
Hawaiian [hɑːˈwaɪɪən] shirt	夏威夷衫
men's coat	男装风衣
gents' coat	男装风衣
women's coat	女装风衣
ladies' coat	女装风衣

children's coat	童风衣
Customs and Excise [ek'saɪz] Office	海关税务机构
Harmonized ['hɑːmənaɪzd] Commodity Description and Coding System (H. S. Code)	协调制商品名称和编码方法
Crocheted ['krəʊʃeɪd]	钩编的
second-hand	二手的，旧的
worn [wɔːn] clothing	旧衣物
fur [fɜː] clothing	毛皮服装
fur lining	毛(皮)的夹里
shell [ʃel]	面料
breeches ['brɪtʃɪz]	马裤
fabric construction	织物结构
composition [kɒmpə'zɪʃən]	成分
compatible [kəm'pætəbl] size	般配的尺码
piping ['paɪpɪŋ]	滚边
morning dress	常礼服
plain	素色的，普通的
cutaway ['kʌtəweɪ]	(从腰至下摆)裁成圆弧后摆的
rounded tails	圆弧后摆
striped [straɪpt] trousers	条纹长裤
tailcoat	燕尾服
imitation [ɪmɪ'teɪʃən] silk lapel	仿丝质驳领
category number	类别号
cotton	棉
wool	毛
man-made fibre	化学纤维
silk blends	真丝混纺
vegetable fibre	植物纤维
duties ['djuːtɪz]	关税
levy ['levɪ]	征收
country of origin ['ɒrɪdʒɪn]	原产国
trade policy	贸易政策
cut-fashioned	裁剪成形
knit-fashioned	编织成形
garment sections	衣片
widening or narrowing	放针或收针
criterion [kraɪ'tɪərɪən] (pl. criteria [kraɪ'tɪərɪə])	准则，标准
World Trade Organization (WTO)	世界贸易组织
authorities [ɔː'θɒrɪtɪz]	当局

statistics [stə'tɪstɪks]	统计学，统计表
statistical [stə'tɪstɪkəl] data	统计数据
regional [riːdʒənəl] regulations	区域性规则
Harmonized Tariff Schedule	美国协调关税制
Outward Processing Program	外加工
Committee for the Implementation of Textile Agreements(CITA)	纺织品协议执行委员会
entry summary form	入境报关总表
knitted shirt	针织翻领衫
collarless	无领的
knitwear	针织服装
sample	样品
working specification sheet (SPEC sheets)	工艺单
army dress blouse	(美)军队制服上装
Hawaiian dress	夏威夷裙
sunray-pleated skirt	百褶裙
skirt suit	裙套
halter ['hɔːltə] top	吊颈(上装)背心
racer ['reɪsə] back vest	窄背背心
spaghetti [spə'getɪ] singlet top	多带式吊带衫
stocking ['stɒkɪŋ]	长袜
overshirt	(可穿在针织衫等外的)宽松式衬衣

Exercises

A. Please use " T " or " F " to indicate whether the following statements are TRUE or FALSE：

(1) The word "jacket" is the general name for almost all short top outer garments.　(　　)

(2) In American English, a "jumper" could mean a sleeveless dress, usually worn over a blouse.

(　　)

(3) The word "shirt" refers to the garment worn by men and the word "blouse" refers only to the garment worn by women.　(　　)

(4) As a countable noun, the word "dress" applies only to women's garments.　(　　)

(5) The word "vest" may refer to a long-sleeved underwear top.　(　　)

(6) In American English, the term pants refers to a lower outer garment and the term underpants refers to underwear bottoms.　(　　)

(7) In the Harmonized Commodity Description and Coding System, knitted or crocheted garments are classified within Chapter 62.　(　　)

(8) The notes for Chapter 62 of H. S. Coding Systems stress that all of the components of a "suit"

must be of the same fabric construction (except for those with piping) , colour and composition; they must also be of the same style and of corresponding or compatible size.

( )

(9) In the U. S. category system, a garment with a category number starting with "3" denotes that the garment is made of cotton. ( )

(10) In international trade, the Customs and Excise Office needs to identify the country of origin for the imported goods so that it can decide what kind of duties should be levied. ( )

B. Please select the word(s) or phrase(s) that makes the statement correct:

(1) _____ refers to clothes worn outside. ( )

    a) "Outwear"   b) "Upper garment"

    c) "Garment top"   d) "Outerwear"

(2) Trousers can be well classified as a _____. ( )

    a) lower garment   b) bottom garment

    c) garment bottom   d) type of bottoms

(3) _____ are underwear for men. ( )

    a) Panties   b) Boxer shorts

    c) Knickers   d) Y-fronts

(4) The upper part and lower part of a suit should _____. ( )

    a) be made from the fabric with the same construction and composition

    b) be of the same style

    c) be of corresponding or compatible size

    d) be in the same colour

(5) According to the United States category system, Category No. 336 refers to garments made from _____ fabric. ( )

    a) cotton   b) silk   c) wool   d) polyester

(6) The correct way to describe in a contract pullovers knitted from a blend of 80% wool and 20% Acrylic yarns is _____. ( )

    a) "Acrylic Pullovers"

    b) "Wool Pullover"

    c) "Pullovers, Wool/Acrylic 80/20"

    d) "Wool/Acrylic Pullovers"

Reading Materials

PFAS, a 'forever chemical' in our apparel
Textile Today Report

PFAS stands for per- and polyfluoroalkyl substances (PFAS), this group of chemicals consists of about 9,000 compounds. This group of chemicals are widely used additives in textile manufacturing due to their unique chemical properties which can make textiles water, stain or heat-resistant. The PFASs are used in our daily articles such as T-shirts, jeans, outdoor jackets, and shoes as a coating to make them water, stain, and heat resistant. So, it makes our life easy, right!

Not completely. Though it comes with many advantages that add special features to our daily textile articles, it has some horrific disadvantages!

PFAS is a toxic chemical. These chemicals are dubbed "forever chemicals" because they do not naturally break down and remain forever in the environment. They accumulate in animals, including humans, and are linked to cancer, birth defects, liver disease, thyroid disease, decreased immunity, hormone disruption, and a range of other serious health problems.

—excerpted from *Textile Today*, December 2021, page 66.

【参考提示】

1. *Textile Today* 是位于孟加拉国达卡的 Amin & Jahan Corporation Ltd 出版的电子期刊 (www. textiletoday. com. bd/emagazine/)。

2. 严格来说，"PFAS stands for per- and polyfluoroalkyl substances (PFAS), this group..." 中 this 前应加 and 且 substance 应是单数形式，何况下文还用到 PFASs 表达。per- and polyfluoroalkyl (即 perfluoroalkyl and polyfluoroalkyl) substance，即"全氟和聚氟烷基物质"。

3. Not completely 应该是 it does not completely makes our life easy 的省略表达。前面的 right 是感叹词。可以看到该处表达偏口语化。

The 7 Best Smart Clothes of 2022
By Brad Stephenson-Updated on January 3, 2022

Smart clothes are traditional clothing items that contain modern technology. Some feature meshed wiring woven into the fabric, while others contain hardware that connects via Bluetooth to an iPhone or Android smartphone. Clothing such as Bluetooth beanies have been around for years, but newer examples of smart clothing expand on technology. Denim jackets can track your Uber's location, and tracksuits can bathe your body in far infrared light. Here are seven of the best smart clothing items available for purchase.

01 Levi's Commuter x Jacquard Jacket

The Commuter x Jacquard is part of a collaboration between Levi's and Google. This non-stretch denim jacket connects to your smartphone via Bluetooth and can screen phone calls, control music volume, and notify you when your rideshare is nearby.

02 Snap Spectacles 3

Snap's third generation of smart glasses boasts improved image quality for videos and photos and dual microphones for better sound recording and faster transfer times. Spectacle 3 claim to " bring augmented reality to life. " The glasses have two HD cameras that capture 3D photos and videos at 60 fps. They have four built-in microphones.

The Spectacles 3 come in two colors: carbon and mineral. Although the cameras on the glasses are still visible, the new design is more subtle, and they don't broadcast to the world that you may be taking snaps.

03 Nadi X Yoga Pants

Nadi X yoga pants can sense when your yoga pose needs refining. Using haptic feedback, the smart pants create small vibrations on the body part you need to adjust.

The Nadi X iOS app offers instructions on how to optimize each pose in addition to proper yoga flows that can be used to curate your own personal yoga class.

Nadi X yoga pants are available for men and women in a variety of sizes. They are machine washable after removing the battery pack that attaches to the rear of your left knee.

04 UA Recover Clothing

UA RECOVER is Under Armour's clothing line that absorbs heat from the human body and then reflects the heat back onto the wearer's skin as far infrared light. This especially useful to athletes because far infrared light encourages better muscle recovery and enhances relaxation.

05 Sensoria Fitness Socks

Sensoria Fitness Socks use advanced textile sensors built into each sock, plus a Core device that snaps into the dock that is attached to the sock to deliver precise data on how your foot lands while walking or running.

The connected app provides tips to improve your walking and running technique, and it tracks your steps, speed, altitude, and distance traveled.

06 Neviano Swimsuits

Designed and made in France, Neviano's swimsuits are stylish and integrated with a UV sensor. The sensor is about half the size of an adult's thumb, waterproof, and connects to the wearer's iOS or Android device. It sends alerts when UV levels are high to remind you to apply more sunscreen.

07 Siren Diabetic Socks

Siren's smart socks, officially referred to as Siren's Diabetic Socks and Foot Monitoring System, use small sensors placed throughout the socks' fabric to measure the foot's temperature at six different points. This data can detect ulcerations which often result in amputations.

—excerpted from https://www. lifewire. com/best-smart-clothes-4176104

【参考提示】

1. Lifewire 是美国 Dotdash 旗下的信息技术网站。

2. Smart clothes 智能服装(不过文中 Snap 公司的眼镜似乎不应归入服装一列)。

3. 品牌类英语专有名词,如无普遍接受的译法则建议直接用原词,如有则不妨使用,如 Bluetooth 蓝牙,因此 Bluetooth beanies 即蓝牙无檐帽。

4. Commuter 是 Levi's 的男装品牌,Jacquard 是 Google 下的智能织物技术项目。

5. HD camera 高清相机。haptic feedback 触觉反馈。Under Armour (UA)美国高端运动装备品牌。ulcerations 溃疡。amputation 截肢。

6. ... snaps into the dock(像揿纽般)揿在(传感器)座上。

文中 more subtle, socks' fabric 和 foot's temperature 等并非英语规范用法。

CHAPTER *2* GARMENT SECTIONS, PATTERNS AND MEASUREMENT LOCATIONS

1 Garment Sections

A garment may have several sections depending on its styling features. Each section may be composed of one or more garment pieces or components made from the same or different fabrics, which are joined together by sewing, fusing, gluing or welding to form a garment.

1.1 Sections in Upper Garments

The simplest garment top or upper garment may have only a body. In other words, it is sleeveless and collarless. Conversely, a relatively complicated style may have many styling features.

1.1.1 Body

The body of most garments has separate front and back garment components joined by side seams. Some knitted garments, for example, some T-shirts, are knitted as tubes and have no side seams. Furthermore, if the garment opens completely at the front, it will have left front and right front. Similarly, the back may be composed of left back and right back components joined by a centre back seam if they are symmetrical. Some garments have a front facing and back neck facing to reinforce them (for example on some ladies' blouse and shirt collars); therefore garment components for the left front facing, right front facing and back neck facing, as appropriate, may be necessary. Styling features of some ladies' garment tops are princess lines. If the princess lines feature in both the back and front together with side seams, then garment components for the left back side, right back side, left front side and right front side will be required (see Fig. 2. 1).

A designer would sometimes use divided seams for styling purposes. Such seams would split the garment component concerned into separate components. For example, if a divided seam occurs across the front body, the front would, as a consequence, be divided into an upper front and a lower front (See Fig. 2. 2).

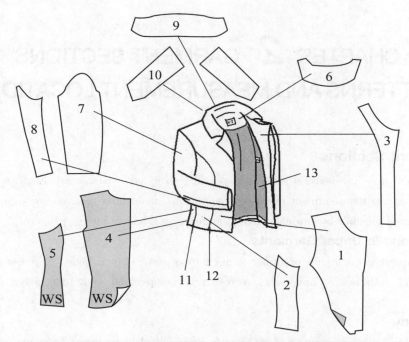

Fig. 2. 1   Illustration of a typical formal jacket showing some of its shell components

1 — right front   2 — right front side   3 — left front facing   4 — right back   5 — right back side

6 — back neck facing   7 — right top sleeve   8 — right under sleeve   9 — top collar   10 — right under collar

11 — side seam   12 — princess line   13 — lining(Remark： "WS" means wrong side.)

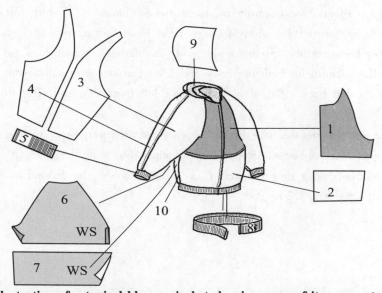

Fig. 2. 2   Illustration of a typical blouson jacket showing some of its garment components

1 — left upper front   2 — left lower front   3 — right front sleeve (raglan sleeve)

4 — right back sleeve (raglan sleeve)   5 — knitted rib cuff   6 — upper back

7 — lower back   8 — knitted rib waistband   9 — right hood   10 — side seam

1.1.2　Sleeves

Many garments are assembled with one-piece sleeves; therefore only two garment components for the left sleeve and right sleeve would be needed. For sleeves with a relatively higher crown, two-piece sleeves with a top sleeve and an under sleeve are usually adopted, in which case, the garment should have the components for a left top sleeve, a right top sleeve, a left under sleeve and a right under sleeve. Raglan sleeves may be composed of front sleeves and back sleeves (see also Fig. 2.2).

A typical shirt sleeve is a one-piece sleeve with a cuff. The number of garment components for the cuffs on a shirt may be either two or four. In the former case, during the sewing operation, each cuff component would be folded with the right side inside, and sewn along both the left and right ends, and then, turned "inside out" (referred to as "bagged out"). In the latter case, two cuff components for one cuff would be put together face to face, and sewn around the sides and the end (see Fig. 2.3). Some designers prefer the latter way, since the dimension of the cuff component is smaller and it helps to increase the fabric utilisation when the marker is made (a description of marker making, which refers to the process of arranging patterns on the fabric to be cut, can be found in Section 1.1 of Chapter 5). The sleeve cuffs on some jackets are formed simply by turning up the sleeve hems. Under such circumstances, no separate cuff components are needed.

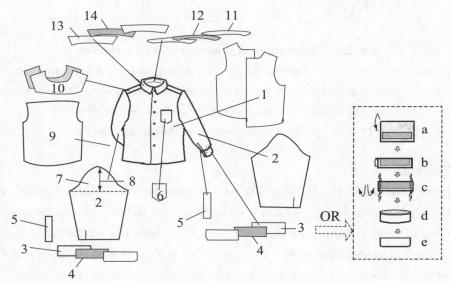

Fig. 2.3　Garment components for a normal shirt

1 — front　2 — sleeve　3 — cuff　4 — interlining for cuff　5 — sleeve vent strip
6 — patch pocket　7 — sleeve crown　8 — crown height　9 — back　10 — yoke
11 — collar stand　12 — interlining for collar stand　13 — collar　14 — interlining for collar
a) applying an interlining at the wrong side　b) folding with the right side inside
c) joining the both sides　d) bagging out with the right side outside
e) the cuff now ready after under pressing

Most sleeves are set-in sleeves, which are sewn into the garment armholes. However, sleeves can also be designed as a grown-on part of the garment body components.

When a garment is designed, various sleeve lengths could be defined according to the desired styling features. Men's garment tops are designed with either full-length or short sleeves. In contrast, ladies' garments could have cup sleeves, elbow length sleeves and three-quarter sleeves in addition to full-length or short sleeves. There are many sleeve style variations, such as the raglan sleeve, Dolman sleeve, batwing sleeve, bell sleeve and bishop sleeve. Furthermore the puff sleeve, melon sleeve, circular sleeve and tulip sleeve are popular short sleeve styles for dresses (see Fig. 2.4).

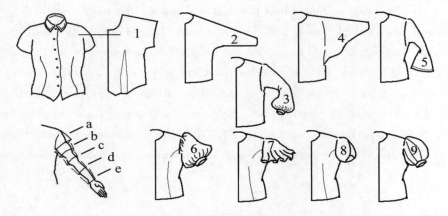

Fig. 2.4 Illustrations of some sleeve style variations

1—grown-on sleeve   2—Dolman sleeve   3—bishop sleeve   4—batwing sleeve
5—bell sleeve   6—puff sleeve   7—circular sleeve   8—tulip sleeve   9—melon sleeve
a) cup sleeve   b) short sleeve   c) elbow length sleeve
d) three-quarter length sleeve   e) full-length sleeve

1.1.3 Collars

There are three basic types of collar. The most common type is the applied collar, which can be further divided into flat collars and band collars. For such collars, garment components for a top collar and an under collar will be necessary. The Eton collar is a typical flat collar, and the mandarin collar is a typical band collar. The shirt collar is actually a combination of a flat collar and a band collar, and it is composed of four garment components viz. a top collar, an under collar, a top collar stand and an under collar stand (see also Fig. 2.3).

The grown-on collar is another type. For the grown-on collar, the under collar is usually part of the body components and the top collar is part of the front facing; consequently there are no separate garment collar components. The roll collar and shawl collar are good examples of grown-on collars.

A garment with a collar conforming to the third basic type has a collar and a rever or lapel. The collar can be made of separate top collar and under collar components and the under collar can be divided into left and right components with diagonal grain lines. The under part of the lapels is part of the front components and the top part of the lapels is part of the front facing (See Fig. 2. 5).

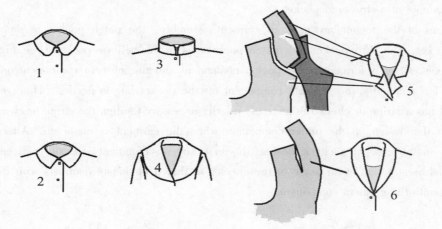

Fig. 2. 5　Illustrations of different styles of collars

1 — peter pan collar (flat collar)　2 — Eton collar　3 — band collar　4 — shawl collar

5 — collar with lapel　6 — roll collar (grown-on collar)

Variations in the collar depth and the collar shape lead to various collar styles. For the collarless garments, styling is achieved through variations in the garment neckline. The boat neck, round neck, V-neck, U-neck and square neck are very common neckline styles (see Fig. 2. 6). Some collarless sportswear has a hood assembled directly onto the neckline and some jackets have a detachable hood behind the collar.

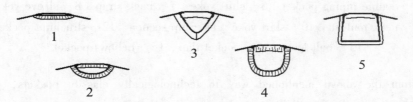

Fig. 2. 6　Illustrations of different styles of necklines

1 — boat neck　2 — round neck　3 — V-neck　4 — U-neck　5 — square neck

1.1.4　Pockets

A pocket requires a pocket opening, and, sometimes, a separate pocket bag. Some pockets have a pocket flap. Technologically, there are also three common types of pockets. These are the insert pocket, patch pocket and structural pocket.

The pocket opening for the insert pocket is cut into the garment component designed to accommodate the pocket. The most common styles of insert pocket are the piping pocket and the welt pocket. Garment components for the piping or welt are also necessary to make up the pocket opening.

The pocket opening for the structural pocket is formed at the edge of the garment component, and no additional garment component is needed for its pocket opening. The seam pocket is one common example of a structural pocket.

In terms of the technology used in garment assembly, the patch pocket is probably the simplest pocket. The shirt pocket is a patch pocket, as is the bellows pocket (see Fig. 2. 7). Since the pocket bag is formed by the pocket component and the garment component onto which the pocket is "patched", no separate component for the pocket bag is needed. However, if the fabric used has a stripe or check design, it is usually necessary to align the stripe or check on the pocket with the design on the under component when the garment is made up. Alternatively, particularly in the case of a check design, because any misalignment is easily noticeable, the pocket is deliberately cut at an angle to the check, so that it obviously contrasts with the design when sewn onto the garment component.

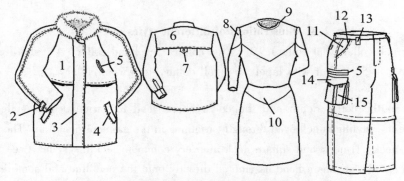

Fig. 2. 7   Pockets, straps and yokes, etc.

1 — coat yoke   2 — sleeve strap   3 — front zipper panel   4 — welt pocket
5 — double piping pocket   6 — shirt yoke   7 — back strap   8 — sleeve yoke
9 — neck rouleau   10 — skirt yoke   11 — hip section   12 — structural pocket
13 — belt loop   14 — pocket flap   15 — bellows pocket

Apart from the above mentioned way to technologically classify pockets, pockets are sometimes named according to their functions, for example, a coin pocket or watch pocket, or according to their positions on the garment, say chest pocket, leg pocket, hip pocket or inner pocket.

1. 2   Sections in Lower Garments

1. 2. 1   Trousers

Usually, four garment components are at least needed to form a pair of trousers, i. e. , a left front, right front, left back and right back. If a waistband is to be attached, and if there are pockets, the corresponding garment components are also necessary. Nowadays most trousers have front openings called flies, if that's the case, the garment components for flies are also needed (see Fig. 2. 8). Some styles of ladies' trousers or slacks have openings at the side seam, usually closed by zips but occasionally by buttons. There should be at least a top fly and an under fly for each side. The number of fly components depends on the style of the trousers, and especially on whether the opening is fastened with buttons or a zipper.

There are also many style variations for trousers that do not require extra components but do affect fabric utilisation. These include plain and pleated fronts, and straight, tapered or flared legs. Other styles do require additional components. For example, the legs of some convertible casual trousers may be divided into two or three zipped-on parts in relation to short, mid-calf length or full-length trousers, and under such circumstances, the front and back components have to be separated. Many trousers have structural front pockets and, if they are not seam pockets, hip sections (see Fig. 2. 7), derived from the front components, would be needed. Turn-ups at the leg hems need no additional components but for Bermuda shorts, leg cuffs may be required.

1. 2. 2　Skirt

Generally speaking, the garment components required for a skirt are much simpler than those for trousers. The simplest skirt is a half circular one without a waistband, for which only one skirt component is necessary. Nowadays many skirts are formed with one component for the front of the skirt and two components for the back (left and right) of the skirt. In such cases, there must be a centre back seam and side seams. For a wrap-around skirt, there should be two front components with the right front wrapping over the left front. In addition to using pleats or slits to provide a larger flare or sweep to the skirt, some designers use gores or godets. As for trousers, garment components for the waistband and pocket may be necessary.

Many style variations for skirts are made on the silhouette. Furthermore, varying the positions of the top of the waistband, changing the skirt length or using yokes can generate different styles (See also Fig. 2. 8).

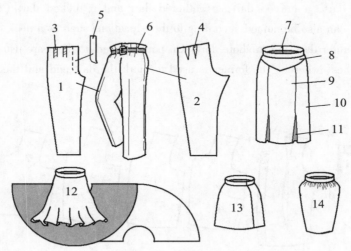

Fig. 2. 8　Illustrations of different styles and features of trousers and skirts

1 — front　2 — back　3 — pleat　4 — dart　5 — fly　6 — belt loop　7 — waistband

8 — yoke　9 — centre front panel　10 — side front panel　11 — godet

12 — circular skirt　13 — wrap-around skirt　14 — peg top skirt

1.3 Other Garment Components and Styling Details

a) **Panel.** A panel is a long narrow garment component usually cut with a lengthwise grain line. Panels are used for a variety of purposes in different garment styles. For example, a front panel can be used to cover a front zipper placket, and a panel would be used to form the cord tunnel at the waist part. A six-panel skirt is formed by six skirt panels (see also Fig. 2. 8, where the skirt on the top right is formed by six skirt panels). This is quite similar to a gored skirt (see Fig. 1. 8). The difference between the "panel" and "gore" in this case is that the widths of panels usually vary but the widths of gores are the same.

b) **Yoke.** Yokes on garment tops are usually around the neck or the shoulder, and they are sometimes designed to emphasise the visual effect of a "strong shoulder". A shirt (dress shirt) yoke usually comprises two components sewn together. Some coats also incorporate yokes to create a more "masculine look". Yokes could also appear at the top part of long sleeves.

Yokes can be found around the waist of some trousers or skirts. They are usually designed to fit closely from the upper hip to the waist and, in some styles of skirts there are unfitted or gathered parts of the skirt hanging from the yokes to create a fuller flared styling effect (see also Fig. 2. 7).

c) **Dart.** A dart is typically, but not always, a V-shaped section marked into the fabric at a particular location such that, when the edges of the V section are sewn together, the flat fabric is transformed into a three-dimensional shape to fit the contours of the three-dimensional human body. The shape, dimensions and location of the dart or darts are determined by a required fit and the location on the body. Consequently, darts may be named according to their shapes such as a wedge dart, a fish dart, a curved dart, a gathered dart and a tucked dart (see Fig. 2. 9). Alternatively, darts can also be named according to their locations such as a neck dart, waist dart, underarm dart, shoulder dart and armhole dart. On paper patterns, darts are usually marked with drill holes and notches, whereby the former is used to mark the dart end and the latter is used to mark the dart sides.

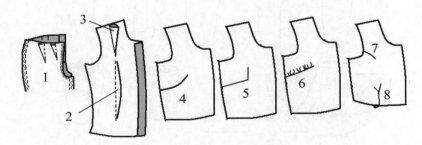

Fig. 2. 9 Illustrations of darts

1 — wedge darts (waist darts) 2 — fish dart (waist dart) 3 — shoulder dart (wedge dart)
4 — curved dart (underarm dart) 5 — biba dart 6 — gathered dart
7 — armhole dart 8 — tucked dart

d) **Pleat, gathers and tucks.** A pleat on a garment is usually formed by folding the garment component at least twice and then stitching one of the ends in a seam. The function of the pleat is to introduce fullness into a garment to allow easier and more movement to the wearer.

There are three types of pleats, that is, the knife pleat, the box pleat and the inverted pleat (see Fig. 2. 10). Pleats are usually located either at the hem of a garment or at the waist in the front of a lower garment such as a skirt or trousers. Multiple parallel short pleats, usually horizontally, are referred to as tucks. All-round parallel pleats on a skirt along its length are called sunray pleats.

Gathers differ from pleats in that the fabric folds in gathers are usually small and irregular, whereas the folds in pleats are regular. Gathers can be used to produce fullness, for example, in puff sleeves. Frills are additional lengths of fabric containing some gathering and they are sewn onto garments to produce a ruffle appearance, and thereby enhance the femininity of a ladies' garment.

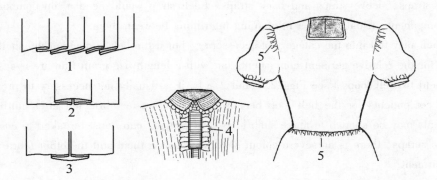

Fig. 2. 10   Illustrations of some pleats and gathers, etc.
1 — knife pleat   2 — box pleat   3 — inverse pleat   4 — frill   5 — gathers

e) **Slits or vents.** Slits or vents are simply cuts into garment components. They are used for various purposes. A slit in a garment hem, at the side or back, can create a larger sweep, whereas slits on a sleeve or leg hem could expand the opening to allow the hands or feet to pass through more easily. Of course, some designers use slits just for styling purposes.

A slit cut into a garment component at locations other than the edge is more often referred to as a vent, particularly if it is intended to allow air to pass through it to enhance the physiological comfort of the wearer. Conversely men's jackets often have single or double vents cut into the backs to create styling features.

Slits on seams usually do not need additional neatening; otherwise, adjustments to the relative garment components, or additional slit facings, or slit strips, etc. are necessary(See Fig. 2. 11).

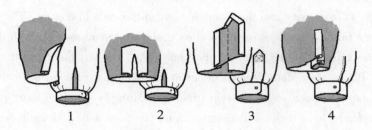

Fig. 2. 11   Some styles of slits

1 — slit on seams   2 — slit with facing   3 — slit with strap   4 — slit with strip

f) **Strap and belt.** Straps are used on some garments often to support the garment on the wearer. For example, some loose-fitting ladies' dresses are designed to have shoulder straps, without which, the dress might fall down. Of course some straps may only have a decorative purpose. Consequently, straps tend to be named according to their locations, for example, shoulder straps, sleeve straps and back straps. Each strap would need a top component and an under component, usually with a reinforcing interlining between them.

A belt may fall into the category of accessories, but if the belt is made of the shell fabric, the patterns for the relative garment components cut with a lengthwise grain line are necessary. If the belt is held by belt loops (see Figs. 2. 7 and 2. 8), it is usually not necessary to prepare special garment components for the belt loop because scraps of waste fabric left after cutting the main components may be enough to make such loops. Of course, care must be taken to ensure that, in using the scraps, there is no severe colour shading between them and the other fabric components in the garment.

2   Patterns

Patterns are the templates used to create the desired shape of the components for the garments. All the patterns for a garment are used to make markers for cutting garment components from fabrics. For lined garments patterns are needed for both the shell fabric and the lining fabric. If some parts of a garment, say the collar and pocket flaps, need interlinings, corresponding patterns are also necessary.

For sample garments, those symmetrical garment components of the same size and same shape may need only one pattern, and in such a case the symmetrical garment component could be traced from the pattern for the other component flipped horizontally. Therefore an upper garment with a centre back seam and side seams will have garment components for the left back and right back, but only the back pattern for one side is required. If a garment has neither a centre front seam nor a front opening placket, only a half front pattern marked with "centre front fold (CFF)" would generally be used to ensure the symmetry of the garment component. This means that the pattern would be placed onto a fabric folded along its length such that the side of the pattern corresponding to the centre front of the garment would be aligned next to the fold in the fabric. Consequently, when the fabric is cut to the shape of the pattern and unfolded, a garment component symmetrical about the fold-line will be created.

When the pattern is prepared, some information about the pattern, such as style number, component name, size of the garment, number of pieces to be cut, must be marked. The words "CBF" or "CFF" should also be marked for symmetrical patterns to be cut on the fold. Notches would be used to identify the seam allowances, balance marks, folding lines and dart sites, and drill holes would be used to mark the dart point and the button locations. Notches should be transferred onto the garment components but in mass production, templates may be specially made to mark the dart points and button locations when sewing the garments.

If the fabric to be used for the shell or lining is woven or knitted, grain lines, relating to the warp direction in woven fabrics or the wale direction in knitted fabrics, must be marked on the pattern so that the grain line on the pattern can be aligned with the fabric length direction when making the marker. Likewise, for woven or knitted interlinings, the grain lines also need to be marked on their patterns.

In a traditional clothing factory without a CAD system, patterns initially produced in paper will be re-cut in cardboard to make more durable patterns, and stored in a pattern room. The patterns may also be manually graded into different sizes for each style. When making the marker, the relevant set of cardboard patterns for the style will be obtained and their outlines traced onto a paper marker. For factories with CAD systems, the data on the shape of the patterns will be input into the computer by digitizing around them and stored in the computer memory. These can then be graded using computer software to generate all the patterns for the required size range. Clearly, if one style of garment has ten patterns for one size and if one garment order for this style comprises ten sizes, the total number of patterns will be one hundred. Using a CAD system to store the pattern data digitally would be much more convenient, and would save a lot of floor space.

3　Measurement Locations

The dimensions of the patterns that need to be produced are clearly related to the relevant anthropometric measurements made on the body that the garment is designed to fit. There are numerous publications on pattern design and cutting and it is, therefore, not the intention of this book to describe it in detail. Chapter 5 gives a brief description of this process. However, it is important at this stage that the reader should be aware of the importance of these measurements, particularly with regard to checking key measurements on the garment during and after its production.

Many measurements need to be made when the clothing is produced and finally inspected to ensure that the garment conforms to the intended size range and that the measurements are within the allowed tolerances.

For a finished garment top, several girths would be checked. These are the bust girth (for ladies' garments, and for men's garments chest girth will be inspected), waist girth, sweep, armhole girth, upper arm girth (sometimes referred to as "muscle" in the U. S. A.) and sleeve opening. Several lengths or widths, such as garment length, sleeve length, across shoulder, across chest, across back, bust or chest width (instead of girth), back neck width, neck depth, cuff depth (if there are cuffs), hem height and the length from nape to waist, may also be measured. If there is a hood, the hood width, hood height from the high point shoulder, hood opening length and hood crown length, etc. may need to be controlled.

For trousers, waist girth, upper hip girth, lower hip girth, thigh, knee girth and leg openings would be checked and the lengths of the outseam, inseam, front rise and back rise are normally measured. Measurements for a skirt are much simpler, and typically these are made at the waist girth, upper hip girth, lower hip girth and the skirt length (usually the CB length).

The size chart (sometimes referred to as the size specifications or measurement chart) will provide full reference, size by size, for an inspection on each style of finished garment. If there is no grading chart, the grading rules could also be derived from the data listed in the size chart.

It must be noted that a workable size chart should specify clearly and precisely the measurement locations, or otherwise errors in measurements may be made. The measurements requested may vary from company to company or from style to style.

For example, the sleeve length could be measured from the shoulder point to sleeve hem, or it could also be measured from the nape (centre back neck) to sleeve hem. The latter is usually used for raglan sleeves. The garment length could be measured along the centre back from the nape to the hem, or at the centre front from the centre front neck to the hem, or it could be measured from the shoulder neck point (or the higher shoulder point abbreviated to HSP, sometimes referred to as the high point shoulder (HPS)—note that the so-called HSP is not necessarily on the shoulder seam.) to the front hem. The armhole could be measured along the curved seam lines of the armhole, but it could be measured diagonally from the end of the shoulder seam at the armhole to the bottom of the armhole at the underarm, seam to seam for garments with sleeves or edge to edge for vests. A curved armhole also makes the exact measurement on the across chest and across back difficult.

Some companies specify the measurements that need to be made by reference to diagrams with the measurement locations clearly marked. Conversely, some companies simply describe these in words such as "waist girth to be measured $16\frac{1}{2}$ inches down from the centre back neck". Fig. 2. 12 shows briefly some commonly used measurement locations and Table 2. 1 gives the location names corresponding to the numbers marked in Fig. 2. 12.

It must be noted that a precise consistent measurement is not feasible in garment production, and therefore, some tolerance must be allowed and specified for each of the measurement locations. In order to avoid subsequent trade disputes, it is advisable to agree a reasonable tolerance during the business negotiation prior to production. This is especially important for companies with newly-established business relations.

The number of measurements specified differs between buyers and, at first sight, the reasons for this can seem confusing. In practice, if patterns for all the sizes are supplied by the buyer and if the corresponding counter samples have been approved, then it should only be necessary to inspect some key locations such as bust girth, waist girth, sweep, sleeve length and the CB length. However, if all the patterns are to be made and graded by the supplier, the buyer might feel the need to check more locations particularly if the buyer did not provide the grading chart and the patterns are to be graded by the supplier. In this case, the buyer should specify, on the size chart, as much information as possible so that the supplier can derive the grading rules from the measurements requested for each size.

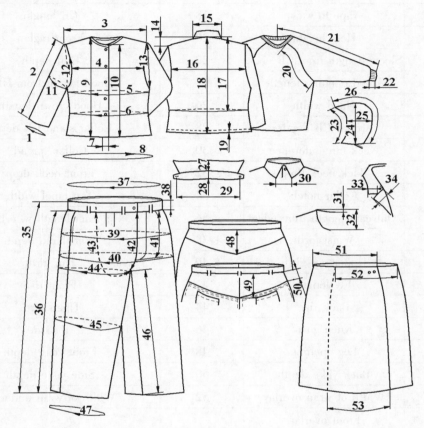

Fig. 2. 12   Examples of common measurement locations

Table2. 1   Names of measurement locations shown in Fig. 2. 12

Number in Fig. 2. 12	Location name	Number in Fig. 2. 12	Location name
1	Sleeve opening	2	Sleeve length
3	Across shoulder	4	Across chest
5	Bust (chest) girth	6	Waist girth
7	Overlap width	8	sweep
9	Garment length from HPS	10	CF length
11	Upper arm girth	12	Armhole curve

(Continued)

Number in Fig. 2. 12	Location name	Number in Fig. 2. 12	Location name
13	Armhole straight	14	Shoulder slope
15	Back neck width	16	Across back
17	Nape to waist	18	CB length
19	Hem height	20	Front raglan
21	Sleeve length from CB neck	22	Cuff depth
23	Hood opening length	24	Hood height from HPS
25	Hood width	26	Hood crown length
27	Collar CB depth	28	Collar stand CB depth
29	Neck circumference	30	Collar spread
31	Back neck drop	32	Front neck drop
33	Collar notch	34	Top lapel width
35	Outseam length from edge	36	Outseam length w/o band
37	Waist girth	38	Waistband depth
39	Upper hip girth	40	Hip girth
41	Fly length	42	Front rise
43	Back rise	44	Thigh girth
45	Knee girth	46	Inseam
47	Leg opening	48	Front yoke depth
49	Back yoke depth	50	Side yoke depth
51	Width of wrap overlay	52	Width of wrap underlay
53	Front overlap		

Words and Phrases

garment component   服装组成部分，衣片

styling feature   款式特征

garment piece   衣片

sewing   缝纫

fusing   热融黏合

welding ['weldɪŋ]   焊接

body   大身

front   前片

back   后片

side seam	（上衣）摆缝，（裤或裙）侧缝，（裤）栋缝
left front	左前片
right front	右前片
placket ['plækɪt]	门襟
left back	左后片
right back	右后片
centre back seam	背中缝
front facing ['feɪsɪŋ]	前挂面
back neck facing	后领圈
princess line	公主缝
left back side	左后侧片
right back side	右后侧片
left front side	左前侧片
right front side	右前侧片
top sleeve	大袖
under sleeve	小袖
top collar ['kɒlə]	面领
under collar	底领
lining	夹里，里子
wrong side（WS）	织物反面
upper front	上前片
lower front	下前片
front sleeve	前袖片
raglan ['ræglən] sleeve	插肩袖
back sleeve	后袖片
knitted rib cuff ['kʌf]	针织罗纹袖口
knitted rib waistband ['weɪstbænd]	针织罗纹腰头
hood [hʊd]	风帽
one-piece sleeve	一片袖
sleeve crown [kraʊn]	袖山
two-piece sleeve	两片袖
right side（RS）	织物正面
bag out	（像翻口袋那样）外翻
fabric utilisation [ˌjuːtɪlaɪˈzeɪʃən]	织物利用率
marker ['mɑːkə]	（衣片样板）排料图，排板图
marker making	（衣片样板）排板，排料
sleeve hem [hem]	袖口边
set-in sleeve	装袖
grown-on [grəʊn-ɒn]	连着的

full-length sleeve	长袖
short sleeve	短袖
cap [kæp] sleeve	帽型袖
elbow ['elbəʊ] length sleeves	中袖
three-quarter sleeve	3/4 长袖
Dolman ['dɒlmən] sleeve	多尔曼袖
batwing ['bætwɪŋ] sleeve	蝙蝠袖
bell sleeve	钟型袖
bishop ['bɪʃəp] sleeve	主教袖
puff [pʌf] sleeve	泡泡袖
melon ['melən] sleeve	瓜型袖
circular ['sɜːkjʊlə] sleeve	披肩短袖
tulip ['tjuːlɪp] sleeve	郁金花式短袖
interlining [ˌɪntə'laɪnɪŋ]	衬头
grown-on sleeve	连身袖
applied [ə'plaɪd] collar	装上去的衣领
Peter Pan collar	彼得·潘领
Eton ['iːtn] collar	伊顿领
mandarin ['mændərɪn] collar	中式立领
grown-on collar	连身衣领
roll collar	翻领
shawl [ʃɔːl] collar	青果领，披肩领
rever [rɪ'vɪə]	驳领
diagonal [daɪ'ægənl]	斜向的
grain [greɪn] line	丝缕线
collar depth [depθ]	领高
neckline ['neklaɪn]	领围线，领圈
boat neck	一字领
V-neck	V 字领（圈）
U-neck	U 字领（圈）
square [skweə] neck	方领（圈）
detachable [dɪ'tætʃəbl] hood	可脱卸风帽
pocket ['pɒkɪt] opening	袋口
pocket bag	袋布，衣袋
pocket flap [flæp]	袋盖
insert [ɪn'sɜːt] pocket	挖袋，镶嵌袋
patch [pætʃ] pocket	贴袋
structural ['strʌktʃərəl] pocket	结构袋
piping ['paɪpɪŋ] pocket	嵌线袋

welt [welt] pocket	贴边袋
piping	滚边，嵌线
welt	袋贴边
bellows ['beləʊz] pocket	风箱袋
stripe [straɪp] or check design	条纹或格子花型
coin [kɒɪn] pocket	零钱袋
watch pocket	表袋
chest [tʃest] pocket	胸袋，(西装)手巾袋
leg pocket	裤腿上的裤袋
hip pocket	臀后的袋
inner pocket	内袋
rouleau [ruːˈləʊ]	滚条
fly [flaɪ]	裤门襟
zip [zɪp]/zipper [ˈzɪpə]	拉链
button [ˈbʌtn]	纽扣
top fly	面襟
under fly	里襟
pleated [pliːtɪd]	打褶的，打裥的
zipped-on	用拉链脱卸的
mid-calf [mɪd-kɑːf] length	至腿肚长的
hip section [ˈsekʃən]	裙或裤的袋垫
leg cuff	裤脚口
half circular skirt	(半圆裙片的)圆裙
pleat [pliːt]	裥
slit [slit]	开衩
sweep [swiːp]	下摆围长
gore [gɔː]	三角形布片
godet [gəʊˈdet]	三角形布片
panel [ˈpænl]	(条形)衣片
cord [kɔːd] tunnel [ˈtʌnl]	绳槽
masculine [ˈmɑːskjʊlɪn] look	阳刚气的外表
dart [dɑːt]	省
contour [ˈkɒntʊə]	轮廓
wedge [wedʒ] dart	契型省
fish dart	鱼型省
curved dart	弧型省
biba [ˈbaɪbə] dart	折角省
gathered dart	带碎裥的省
tucked dart	带褶子的省

37

neck dart	领圈处的省
waist dart	腰部的省
underarm [ˈʌndərɑːm] dart	腋下的省
shoulder dart	肩部的省
armhole dart	袖窿处的省
dart end	省尖
dart sides	省端
notch	刀眼
stitching	缝合
seam	缝子
knife pleat	刀裥
box pleat	箱式裥
inverted pleat	阴裥
tucks	(平行的)裥
gathers	碎裥
frill	褶边
neatening [niːtnɪŋ]	毛边收光，毛边处理
slit facing	开衩处贴边
slit strip [strɪp]	开衩处的滚条
shoulder strap [stræp]	开衩处的镶片
sleeve strap	袖襻
back strap	背襻
belt	腰带
accessory [ækˈsesərɪ]	辅料
belt loop	蚂蟥襻
scraps [skræps] of fabrics	织物碎片，碎料
colour shading	色差
pattern [ˈpætən]	衣片样板
template [ˈtemplɪt]	模板
lined garment	有夹里的服装
paper pattern	纸样
lining	里料
symmetrical [sɪˈmetrɪkəl]	对称的
trace [treɪs]	描，描绘
flip [flɪp]	翻转
centre front fold (CFF)	沿前中线折，无前中缝
front opening placket	前门襟
centre front (CF)	前中线
style number	款号

piece name	衣片名称
centre back fold（CBF）	沿背中线折，无背中缝
seam allowance [ə'lauəns]	缝头
balance mark	对位记号
folding line	折边线
dart point	省尖
button location	纽扣位置
warp [wɔːp]	经纱
wale [weɪl]	(针织物)纵行
CAD system	计算机辅助设计系统
to be manually ['mænjuəlɪ] graded	待手工推档
input	输入
digitising ['dɪdʒɪtaɪzɪŋ]	通过数码转换板输入
computer memory	计算机储存
software ['sɒftweə]	软件
garment order	服装订单
floor space	占地空间
anthropometric [ˌænθrəupəu'metrɪk]	人体测量学的
measurement	测量，测量尺寸
measurement location	测量部位
bust [bʌst]（girth [gɜːθ]）	女装胸围(围长)
chest（girth）	(男装)胸围(围长)
waist（girth）	腰围(围长)
armhole（girth）	袖窿(围长)
upper arm（girth）	上臂围(围长)
sleeve opening	袖口大
garment length	衣长
sleeve length	袖长
across shoulder	肩宽
across chest	胸宽
across back	背宽
bust or chest width	半胸围
back neck width	后领宽
neck depth	领深
cuff depth	袖口(克夫)宽
hem height	下摆宽(折边高)
(the length) from nape [neɪp] to waist	背长
hood width	帽宽
hood height	帽高

high point shoulder（HPS）	高肩点
hood opening length	帽开口长
hood crown length	帽顶长
upper hip（girth）	上臀围(围长)
lower hip（girth）	臀围(围长)
thigh	横裆
knee（girth）	膝围(围长)
leg opening	裤脚口大
outseam ['aʊtˌsiːm]	裤长
inseam ['ɪnsiːm]	裤脚长，裤脚内侧缝
front rise	前直裆
back rise	后直裆
skirt length	裙长
CB length	背中线长
size chart	尺寸表
measurement chart	尺寸表
size specifications	尺寸表
grading chart	推档表
grading rule	推档变量
shoulder point	肩点
nape（centre back neck）	后颈点(领圈背中线处)
centre front neck	领圈前中线处
hem	下摆
shoulder neck point	肩颈点
higher shoulder point（HSP）	高肩点
overlap ['əʊvə'læp]（width）	叠门(宽)
armhole straight	直量袖窿，挂肩
armhole curve	弯量袖窿
CF length	沿前中线衣长
shoulder slope	肩斜
front raglan	插肩袖前挂肩
collar notch [nɒtʃ]	驳领缺嘴宽
neck circumference [səˈkʌmfərəns]	颈围
collar CB depth	领子背中线处高
collar stand CB depth	领脚背中线处高
collar spread [spred]	领尖间距
back neck drop	后颈深
front neck drop	前颈深
top lapel width	驳头宽

w/o (without)	没有，不计，不含
waistband depth	腰头宽
fly length	裤门襟长
thigh [θaɪ] girth	大腿围长，横裆
front yoke depth	拼腰前深
back yoke depth	拼腰后深
side yoke depth	拼腰侧深
width of wrap overlay ['əʊvəleɪ]	上叠片宽
width of wrap underlay ['ʌndə'leɪ]	下叠片宽
front overlap	前叠门

Exercises

A. Please use "T" or "F" to indicate whether the following statements are TRUE or FALSE：

(1) A typical shirt collar is composed of four garment pieces, that is, a top collar, an under collar, a top collar stand and an under collar stand.　　　　　　　　　　(　　)

(2) The welt pocket is a typical example of an inserted pockets.　　　　　　　(　　)

(3) The difference between a "panel" and a "gore" for skirts is that the widths of panels usually vary but those of gores are the same.　　　　　　　　　　　　　　(　　)

(4) The function of darts is to transform flat fabric into a three-dimensional shape to fit the contours of the human body.　　　　　　　　　　　　　　　　　(　　)

(5) The so-called HSP is always located on the shoulder seam.　　　　　　(　　)

(6) The "across chest" and "across back" measurement should be checked for every finished garment.　　　　　　　　　　　　　　　　　　　　　　　(　　)

(7) The measurement on the armhole must be made along the curved seam lines of the armhole.

　　　　　　　　　　　　　　　　　　　　　　　　　　　　　　　(　　)

(8) The sleeve length of all garments must be measured carefully from the nape to the sleeve hem.　　　　　　　　　　　　　　　　　　　　　　　　　(　　)

(9) It is not reasonable to ask a garment seller to supply garments with an exact figure of bust measurement without tolerance.　　　　　　　　　　　　　　　(　　)

(10) Buyers should specify, on the size chart, as much information as possible if the grading rules need to be derived by the supplier.　　　　　　　　　　　　(　　)

B. Please select the word(s) or phrase(s) that makes the statement correct：

(1) A typical shirt sleeve _____.　　　　　　　　　　　　　　　(　　)

　　a) is a raglan sleeve

　　b) comprises a top sleeve and an under sleeve

　　c) is a one-piece sleeve

d) has a cuff

(2) Which of the following pieces is the component(s) in a typical shirt?　　　　　(　　)

　　a) Yoke　　　　　　　　　　b) Back neck facing

　　c) Left front　　　　　　　　d) Left back side

(3) Which of the following pockets falls under the category of patch pocket?　　(　　)

　　a) A double piping pocket at the hip

　　b) A typical shirt pocket

　　c) A bellows pocket

　　d) A seam pocket

(4) Notches would be used on a pattern to identify _____.　　　　　　　(　　)

　　a) seam allowances　　　　　b) dart sites

　　c) balance marks　　　　　　d) folding lines

(5) The measurement relating to the length of a pair of trousers is the _____.　(　　)

　　a) hip girth　　　　　　　　b) inseam

　　c) front rise and back rise　　d) outseam

C. Please list the names of the garment pieces for all the shell components for the garment illustrated below:

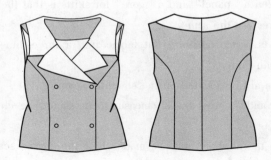

Reading Materials

Pattern Review: Vogue 1786

By Anna Mazur

(Vogue 1786, view B, in silk crepe de Chine plissé from MoodFabrics. com)

This is a versatile shirt with a trapeze silhouette. Design elements include: a traditional shirt collar with stand; hidden front-button placket; deep back yoke; semicircular lower back section; curved hemline that is lower at the back than the front; a self-fabric belt; and cuffed sleeves with curved seams at the elbow, which create a lantern silhouette. The sewing instructions are easy to follow with enough match points to ensure accuracy and ease of assembly. This design offers opportunities to make it your own. Combine sheer fabrics with solids or larger with smaller scale prints. Play with combinations for the lower sleeve and back yoke pieces. The front and back sections are large enough blank canvases to add embroidery or iron-on crystals. Possible areas to add beading and trims are on the front placket, cuffs, and back yoke. Our seamstress says:

　·Be careful when cutting, as the printed cutting line styles are not the same on each pattern piece.

　·Make a muslin before cutting your fashion fabric, as there is a lot of ease in this pattern.

Recommended fabrics are cotton shirting, linen, and poplin. In addition, lace, cut velvet, silk satin, voile, lawn, and lightweight denim are suitable. In a fancier fabric, the shirt works as a jacket over a slip dress or with a camisole and slim pants.

—excerpted from *Threads* #216, Winter 2021, page 32.

【参考提示】

1. *Threads* 是位于美国康涅狄克州的 Taunton Press, Inc. 公司出版的专门讨论与介绍服装面料、服装缝制等内容的双月刊杂志(www. threadsmagazine. com/magazine)。

2. 有一百多年历史的 Vogue pattern 样板设计公司的 Vogue 是美国四大(Big 4)样板品牌之一，这是关于它的 Vogue 1786 样板的评述，view B 应是当时一起发布的 B 号视图。该公司先后被并购加入 McCall、CSS Industries 和 Design Group 旗下。

3. crepe de Chine 双绉。plissé 指的是"泡泡纱"或"产生泡泡纱效应的整理"。

4. trapeze silhouette "梯形"造型。这里的 trapeze 舶自法语 trapèze，相当于英语 trapezoid。该造型服装从肩部、颈部至胸围线收身，然后逐渐张开而形成很宽松的下摆。

5. ensure accuracy and ease of assembly 这里的 ease 和 accuracy 并列说明 assembly 的，不作通常的"宽放量"解释。

6. solids = solid fabrics。… or larger with smaller scale prints 即 … or combine larger with smaller scale prints(……或较大花型的印花布和较小花型的印花布组合)。

7. blank canvases 空白的画布。

8. Make a muslin before cutting your fashion fabric 意思是"在裁剪做时装的面料前先用普通棉平纹细布打个样"。

9. cut velvet 割绒布。Voile 玻璃纱。

Pyjama Season

As nights get colder, it's all about making the coziest pyjamas. We look at three patterns that will keep you and your family toasty warm right through to the spring!

Carolyn Pajamas by Closet Core

Designed for chic lounging, the Carolyn Pajamas are tailored with a modern, figure-flattering cut. The top has a classic notched collar, curved hem and breast pocket with three sleeve options. For a matching pajama set, choose between a straight-legged trouser (with or without cuff) or cuffed shorts. Both have a comfortable elasticated waist, pockets and a faux fly. Personalise this pattern with contrasting piping details for an elegant finish. This pattern comes in sizes 0−20 and can be made in cotton flannels, linens, quilting cottons, shirting and lawn.

Unisex Pajama Pants by Wardrobe By Me

This simple pair of pajama pants has an elastic and pull string in the waist casing. As they don't have a side seam, they're super comfortable for sleeping and have several lengths to choose from. This pattern is very fast to make and also ideal for beginners and has unisex sizing. Choose fabrics that will be comfortable in bed like cotton, viscose, silk, polyester, or a blend of other fibres.

Footed Pajamas by Jalie Patterns

This all-round pattern makes fun gifts for loved ones who are always cold at night! It includes a footed pair of pajamas or with ribbed ankle cuffs. There's a centre front zip (inseam for babies) plus a practical side pocket and ribbed collar and sleeve cuffs. The pattern starts at size 12M for the little ones and has 29 sizes so you can make one for the whole family — girls and boys, women and men as well as plus sizes. Fleece with 30% stretch across the grain with some stretch in the length is ideal.

— excerpted from *Modern Sewing starts here*, Edition 19-December 2021, page 11.

【参考提示】

1. *Modern Sewing starts here* 是位于英国的面料、服装样板等方面的批发及分销商 Hantex (hantexonline. co. uk)发布的电子期刊。该公司发布并提供欧洲的许多独立样板设计师或公司的大量作品。可以注意到，某些非英语母语的设计师在发布的作品中英语表述不够传统规范，也可以看到文中有英式和美式英语拼法的混杂。

2. Closet Core，Wardrobe By Me，Jalie Patterns 都是通过 Hantex 公司发布样板作品的顶尖品牌的独立样板设计师或机构。

3. Pyjamas 睡衣。该词有时缩略为 Pjs。Pyjamas 和 pajamas 都是"睡衣"，前者为英式拼法，后者是美式拼法。它们的单数形式在规范英语中只用作定语，如 pajama bottoms，即"睡裤"。

4. Carolyn Pajamas 是以 Carolyn 命名的一款睡衣，通常为上下套装，平驳领，比较收身，用镶色滚条镶边。如作翻译练习，不妨直译为"Carolyn 款睡衣"。

5. figure-flattering 指"显得身材更好的"，"更修身的"。notched collar 平驳领。

6. a straight-legged trouser (with or without cuff) 带或不带翻裤脚的直筒裤。显然传统规范英语中不用 a ... trouser 的表达而用 "straight-legged trousers"。单数形式 trouser 一般用作定语，如 trouser pocket。

7. quilting cottons 这里应该指"加上绗缝(且常夹有填料)的棉布"。quilt 指"绗缝"，cottons 指"棉布"。Shirting 和 lawn 分别指"细平布"和"上等细布"。英语中有时习惯会用"某类服装名称 + ing"表示做这类服装的"料子"。如 shirting——做衬衫的料子，sheeting——做床单的料子。可能因早年人们用细平布做衬衫，用粗平布做床单，所以有了如今 shirting——细平布，sheeting——粗平布的说法。其他如 skirting——裙料，coating——大衣料，panting——裤料等都是如此。

8. faux fly 假裤襟。footed pajamas 连脚睡衣。plus size 超大码。Fleece 起绒布。

CHAPTER *3*　GARMENT FABRICS, ACCESSORIES, LABELS AND HANGTAGS

1　Garment Fabrics

Garments can be made up by garment components cut from fabrics according to the patterns, or from shaped panels knitted with yarns through the techniques of narrowing and widening on flat weft knitting machines. Modern technology has made it possible to knit a seamless garment directly on fully-fashioned V-bed machines and some circular weft knitting machines.

Cut-fashioned garments may be either lined or unlined. Lined garments need both an outer shell fabric and the lining, which may be either knitted or woven. Non-woven fabrics may also be used as shell fabrics for disposable garments such as surgical gowns, protective clothing or underwear for travellers or incontinent patients. Shell fabrics may be reinforced with interlinings to enhance the performance, drape or appearance of the garment. Since interlinings are normally not visible, the emphasis in their selection is cost and performance, and in this respect, non-woven fabrics tend to be preferred to woven and knitted fabrics for use as interlinings.

1.1　Shell Fabrics

Although there are some garments made of leathers or furs, most garments are made from textile materials. Many types of fabrics may be used for the shell. These range from light weight fabrics such as a silk habotai of around 40 grams per square metre to heavy weight fabrics such as a woolen Melton of about 700 grams per square metre, and are produced from natural fibres, man-made materials or blends of both. The final price of a garment depends to a great extent on the material used and the quality of the shell fabrics. Colours, textures and designs knitted, woven or printed on the shell fabrics would be the main factors that make garments attractive to customers. Without doubt, a garment made from luxurious fabrics in beautiful eye-catching colours with an elegant style would be regarded as a high-class item of apparel.

Many factors have to be taken into account when choosing a suitable shell fabric. For example, is it for underwear or outerwear? What climate is the garment to be worn in? Is the garment to be worn on special occasions? What is the intended wearer's social status? All these questions would have to be answered before the fabrics are finally chosen.

Fabrics for shell fabrics need to be both functional and fashionable. In addition, fabrics worn next to the skin need to be particularly comfortable to wear. They should invariably be soft, non-irritating, and possibly absorbent and, for women, they may be designed to enhance their feeling of femininity. Cotton and silk are the predominant natural fibres used for underwear; and polyester, nylon and elastomeric yarns (such as Lycra^{TM} and Spandex^{TM}) are the most commonly used synthetic materials.

For underwear, weft knitted fabrics offer good porosity and extensibility. Plain weft knit (single jersey) is widely used, but interlock may offer better warmth retention. Other double jersey rib structures, such as 1×1 or 2×2 ribs, are more extensible and, therefore, are particularly suitable for cuffs and necks to ensure snug-fitting. There are some disadvantages for knitted fabrics. Their relatively poor shape retention and higher shrinkage are typical problems that they present. Furthermore, edge or end curling may cause problems during making-up when the knitted fabrics are cut. Fabrics woven from silk, cotton or nylon usually in plain woven structures, such as silk habotai, crepe de chine, taffeta and cotton sheeting or shirting, may also be used for underwear. Tricot warp knitted fabrics, knitted from polyester or nylon, and lace are also popular for use in lingerie.

Fabrics for outer garments can be chosen from a much wider range of materials, depending on the intended usage. 100% cotton fabrics could be used for casual clothing and all wool fabrics could be used for formal garments. 100% synthetic materials or blends with, usually, natural fibres could enhance fabric properties in terms of stiffness, abrasion-resistance and crease-resistance, etc.

Woven shell fabrics are widely used for outer garments and plain woven structures, such as poplin, chintz and palace, tend to be the most popular. Some casual jackets or trousers are made of khaki, and some uniforms made of gabardine or serge. They are commonly-used twill structures. Yarn dyed or printed checks and stripes are classical patterns used in woven fabrics for outerwear. Floral patterns are popular patterns for dresses. Knitted fabrics can also be used as shell fabrics mainly due to their higher productivity, particularly for applications where extensibility, warmth and closeness of fit are required. However, their relatively poor shape retention and a propensity to pill (a phenomenon in which small "pills" or balls of fibres on the fabric surface are caused by entanglement of fibres when fabrics are washed, dry cleaned or rubbed) preclude knitted fabrics from overwhelming the woven structures as the shell fabrics.

1.2 Linings

There are two types of linings. One is sewn into the garment and the other is detachable, so that it may easily be removed from the shell to facilitate washing of the shell. Some garments could have both types of linings.

One of the reasons for incorporating a lining into a garment is to make it warmer. In addition to providing warmth, another function of sewn-in linings is to let the wearer's hands slide down the sleeves easily. For such a purpose, smooth fabrics such as polyester taffeta, polyester pongee, nylon taffeta and rayon satin are widely used. For winter clothing, the silky feeling lining of man-made filaments would have an uncomfortable cool touch; therefore, sometimes, especially for children's winter coats or jackets, woven cotton sheeting or cotton flannel or plain knit, cotton or viscose, may be used. For some nylon casual sportswear, warp knitted meshes may be used as the lining material to prevent the nylon from clinging to the body.

Warp knitted fabrics knitted from polyester or nylon multiple continuous filaments into Locknit or Reverse Locknit structures were used as the lining for some workwear because of the smooth surface and the lower cost.

The detachable lining is usually buttoned-on or zipped-on. It could be made from weft knitted pile fabrics, but with down filling or polyester wadding, etc., the detachable lining provides more warmth to wearers. The detachable linings of some garments are made of fur or the like, and if so, the garments would fall into the category of Chapter 43 in the Harmonized Coding System instead of Chapter 61 or 62, even if the shell fabric and any sewn-on linings are textile materials.

2　Accessories

Accessories or trimmings are necessary for almost all garments. It is important to know how the accessories should be selected and the relationship between the accessories and the garment production. For metal accessories or metal-coated accessories, it is essential to realize that most countries have issued special regulations on the use of lead or nickel since they might be harmful to the wearer's health. For accessories to be used for children's garments, they must not contain toxic elements and must withstand a force of not less than 90 Newtons, as typically specified by buyers, before they are pulled off a garment.

2.1　Sewing Threads

Except for very few items of clothing, most garment components are sewn together using sewing threads. The main functions of the threads in clothing production are joining, edge neatening, topstitching and quilting. Generally speaking, threads to be used should match the fabrics on which the seams are to be formed and the quality of the threads should meet adequate performance requirements.

Sewing threads are either staple spun or continuous filament or a combination of both. Core-spun sewing threads are some of the most commonly-used sewing threads. They comprise a continuous multifilament core to provide adequate strength and a cotton covering to protect the core from the high sewing temperature caused by the friction between the threads and the sewing elements that would otherwise melt the filaments. Traditionally, staple spun sewing threads were spun from cotton, typically 3-ply (3 singles cotton yarns twisted together). Nowadays, polyester thread is the most commonly-used staple spun sewing thread.

Continuous filament nylon or polyester sewing threads have high tenacity and abrasion resistance and are used for applications where there is a particular requirement. Bulked or textured continuous nylon multifilament has a high extensibility and good elastic recovery and, therefore, is used for knitwear and swimwear, etc. Its softness makes it well suited for overedge stitching for garments worn next to the skin. Monofilament threads are also used for some applications. Special threads spun from Nomex are used in potential high temperature applications such as fire-fighting clothing and racing drivers' suits.

The ticket number of a sewing thread defines its size. For cotton staple spun sewing threads, the ticket number is defined as three times the resultant cotton count for the sewing thread. So, for example, a 2 fold 40^s cotton sewing thread would have a resultant cotton count of 20; therefore its ticket number would be 60 (3 times 20). For synthetic sewing threads, a ticket number based on metric count, tex or denier is used. It is important to be aware of the method of thread sizing used when a thread specification is made.

For staple spun yarns in particular, the twist inserted is critical to provide the thread with suitable strength and flexibility; if the twist inserted is excessive, it could cause the thread to snarl or form knots or loops.

The word "joining" means to assemble one garment piece to another; however a garment piece may be joined to itself as happens when preparing a tie. The thread for joining should therefore have sufficient strength, and furthermore, it should not shrink more than that of the garment components to be joined when laundered or dry cleaned.

Edge neatening stitching is produced by sewing over the edge of the fabric using an overlock stitch. The stitch can be used for both woven and knitted fabrics in this respect. The construction of the stitch is that, as well as covering the raw edges so as to prevent them from fraying, it is highly extensible, which makes it ideally suited for sewing knitwear which is inherently more extensible. The looper threads within the stitch are often textured multifilaments, which provide the strength, extensibility and, for underwear and swimwear, the softness required.

Topstitching is used to impart additional strength to the seams joining the garment pieces, and, since it is also a visible stitch, it also offers a decorative element to the garment. A typical application is on shirt collars. The thread for topstitching should not have a bigger shrinkage than that of the garment pieces stitched and it should either match the ground colour, tone in tone, or provide a distinctive contrasting colour. In any event it must have good colour fastness.

One of the functions for quilting is to hold the wadding or filling in a garment in position. Nowadays some designers have moved away from using the simplistic vertical, horizontal or check quilt patterns, because computer-controlled quilting machines now can produce very delicate quilt patterns. The strength of the quilting threads should be taken into account, especially for heavy wadding, and the thread shrinkage must be controlled. The colour of the thread should be that specified on the working sheets and its colour fastness should also be good.

Other threads are used for fastening buttons, making buttonholes and for embroidering. Carefully choosing the threads according to the specifications given by the designer is most important.

In international trade, the garment importers could specify the threads to be used in the working sheets or specification sheets for garments. The particulars about the threads must be carefully considered to determine the type and linear density of thread that should be used. Some buyers may designate a certain international thread brand. However, consideration should be given to the availability of such threads. To import thread would cost more, both in time and money. If possible, the garment supplier can discuss with the buyer whether the thread could be substituted with a local brand. Thread consumptions are sometimes given in the working sheets, but they are generally given only as a reference. The garment manufacturer can often estimate thread consumption according to experience when making the quotation and such consumption can be verified during the garment sample making.

2.2　Fasteners

Openings such as front openings and some pocket openings need fasteners. Nowadays various fasteners may be found in the market and some of them are used not only for fastening but also for decorative purposes.

2.2.1　Buttons

Buttons are the simplest fasteners used today. If buyers supply working sheets, the style number or article number, size, quantity and colour of the buttons to be used will appear on the working sheets. However, the actual shapes and colours have to be specified by samples.

Metal buttons or buttons made of leather tend to be expensive and, therefore, plastic buttons are often specified. Metal-coated buttons and leather-imitation buttons offer similar effects but cost much less. Normal coloured buttons are cheaper than those with special colours such as pearlescent or colours with marble grains.

Buttons in special shapes will need special moulds or dies. If the garment contract specifies a large quantity, the cost of producing the mould for them may be negligible. However, if the quantity is very small, the cost of each button would be significant. As a garment supplier, patient negotiation with your counterpart to agree a specification can invariably find a mutually beneficial solution. The buyer might be persuaded to increase the quantity within the garment orders, or replace the button, or accept higher garment prices.

Attention must be paid to specification for the buttons and the available clothing manufacturing technology. Some clothing factories have button sewing machines for two-hole or four-hole buttons but they might not have the machines for shank buttons; some factories may have straight buttonhole machines, but they might not have an eyelet buttonhole machine. If such factories are designated in the garment contract as the manufacturer and if such buttons or buttonholes are specified in the garment orders, the supplier may find it difficult to fulfill the order unless these operations are allowed to be sub-contracted to a manufacturer who has the necessary equipment. For wrapped-around buttons, the shell fabrics will be used to make the buttons. However, for some types of fabrics, the colour shading might not be properly controlled; therefore, the feasibilities of using the wrapped-around buttons with such fabrics must be taken into consideration. Furthermore, buttons on the outside of a garment generally need one substitute button or spare button, and even if it is not specified, it should always be confirmed with the buyer whether or not the substitute button is necessary.

The size of a button is usually defined by the unit of ligne. A ligne is 1/40th of an inch, or 0.635mm. Therefore, a button with a diameter of 5/8ths of an inch would be specified as 25 ligne ($40/8 \times 5$). In some countries, the size of a button is expressed by its diameter in millimeters.

2.2.2　Zippers

Zippers or zips are also important fasteners for garments. The basic components of a zipper are the zipper teeth and the slider. The zipper teeth are lined up along the edges of two zipper tapes and the slider is composed of a slider body and a pull-tab, or gripper, some of which have a pin lock to prevent the zipper from opening inadvertently. When the tab is pulled, the slider will move along the teeth, causing the two zipper tapes close or open. There could be a pendant on the gripper to add some aesthetic element. Some zippers have double-tag sliders; some have two sliders and thus the zipper could be opened or closed in both directions.

There are three types of zippers according to the materials used for the zipper teeth. Metal zippers have copper or aluminum teeth, and therefore, they are usually more expensive. They usually require a zipper guard or facing to prevent the zipper from being in direct contact with the skin. The second type is the nylon zipper, which has teeth formed with a spiral of nylon monofilament coils; consequently, such zippers may be referred to as "spiral zippers" and they are usually the cheapest of the three. The last type of zipper is the plastic zipper with moulded plastic teeth, and the so-called Vislon zipper falls into this type.

Zippers could also be classified as either open or closed zippers. Open zippers are used for fastening the front opening and attaching a detachable lining or hood. The zipper tapes of an open zipper can be fully separated when the slider is moved to the bottom end. The slider is then retained on one of the zipper tapes, at the bottom end of which there is a retaining box used to contain the insertion pin at the bottom end of the other zipper tape. It is very important to make clear whether the slider is on the left hand side of the zipper tape or on the right hand side, especially for the zippers used for detachable hoods. The closed zippers cannot be fully opened due to the bottom stop at the bottom of the zipper. They may be used for zipped flies or for fastening pocket openings or neck openings. Generally, moulded top or bottom stops are required. For zippers that are supplied on a roll which must, therefore, be cut to size, the stops are applied during garment production. In this case, metal claw (prong) stops would have to be used, but attention must be paid to ensure that the stops do not have rough or sharp edges, and a zipper guard or facing should be used to prevent zipper stops from being in direct contact with the skin.

For ladies' skirts or trousers, "invisible" or concealed zippers could be used. They are usually of the closed type and only the slider appears after the zipper is closed. "Invisible" or concealed zippers are not used on garments for children 3 years and under.

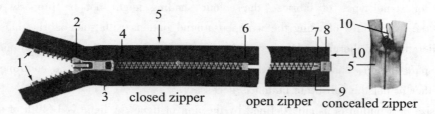

Fig. 3. 1   Zipper elements

1 — top stop   2 — slider   3 — pull tab   4 — teeth   5 — tape   6 — bottom stop
7 — box pin   8 — retaining box   9 — insertion pin   10 — reinforcement tape

For spiral zippers or plastic zippers, the colours of the tape, teeth and the slider are usually tone in tone; for metal zippers, the designer should specify the tape colour as well as the metallic colour of the teeth and the slider. If there is a pendant, its shape and colour should be specified by a sample.

The size of a zipper is specified by its length and the zipper number. The higher the zipper number, the thicker the teeth are and the bigger fastening strength that will be produced. Zipper Nos. 4 and 5 are usually used for outerwear, and No. 2 is used for garments made from fine and light fabrics. For zippers used for front openings and detachable hoods or linings for adults' garments, the zipper length is generally not varied from size to size with the garments. However, the zipper length used at the front openings for children's coats or jackets would vary because the garment lengths for children whose ages range from 8 to 15 will be quite different. Of course, it is not necessary to specify a different zipper length for each size. The usual approach is to divide the sizes into two or three groups, and for each group, zippers with the same length will be used. If such information is not specified in the working sheets, the supplier should ask the buyer to clarify it.

2.2.3　Snaps

It is believed that snaps or press buttons were introduced into garments only in 1930s. These fasteners have proven very popular because of the easy way in which they fasten by "snapping" or "popping" (which gives snaps the nickname of "popper") together. Snaps are usually expensive compared to plain buttons, because the main parts of the snaps are usually made from metal. Interlinings are normally used to reinforce the fabric at the position where the snaps are to be set.

There are two basic types of snaps, according to the way they are fixed on garments. One is fixed with prongs and the other with a cap for the socket and a post for the stud (see Fig. 3.2). For knitted fabrics the prong type should be used and an interlining should be applied. The post type is generally used on heavy weight or tightly-woven fabrics.

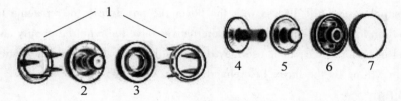

Fig. 3.2　Components of snaps

1 — prong　2&5 — stud　3&6 — socket　4 — post　7 — cap

2.2.4　Velcro

Velcro is the trademark used for a kind of tape fastening, which is composed of a strip of nylon tape with a surface of tiny hooks and a strip of nylon tape with a surface of very soft loops. When the hook part is pressed onto the loop part, the hooks will be caught by the loops, and thus both tapes will be fastened together quite securely and be able to withstand significant forces before they separate. However, if the hook strip is peeled from the loop strip, the two strips can be separated quite easily. The Velcro fastener is easy to use like the snaps but its non-metal structure makes it much cheaper. However, it must be noted that Velcro has several disadvantages. It tends to accumulate hair and dust in its hooks, and the stiff hooks on the Velcro might accidentally snag loosely woven or knitted fabrics or damage drawstrings braided with nylon multifilament. Furthermore, some people dislike the tearing noise made when unfastening Velcro.

2.3　Interlinings

One of the main functions of an interlining is to reinforce the garment part where it is applied. The interlinings used together with snaps, rivets or eyelets serve such a purpose. The other reasons to use interlinings are to increase the stiffness of the garment pieces or to stabilize the shapes of garment sections.

Interlinings with woven or knitted substrates perform well and are generally durable in use. Knitted interlinings are more likely to be applied to fabrics with extensibility or elasticity. Since the grain lines of such interlinings must follow those on the garment pieces, relatively lower utilisation would be inevitable. On the other hand, most interlinings with non-woven substrates are almost "isotropic" since the fibres in the substrates are randomly orientated, and thus no grain lines need matching. This could greatly reduce the wastage in cutting. In mass production, non-woven interlinings are widely used because of their lower cost and higher utilisation but their disadvantage of poor abrasion resistance preclude them from being used in applications where good abrasion resistance is needed.

Although interlinings could be sewn onto the garment pieces, in the main, fusible interlinings are used. These have resins printed, coated or sprayed onto the fabric substrate and they are attached to the shell fabric by the application of heat and pressure (fusing) using flat bed or conveyorised presses or, for small scale production, by steam irons.

If the working sheets from the buyer show that interlinings of certain foreign brands are needed, the supplier can still discuss with the buyer the possibilities of replacing the designated brand with certain local interlinings. If local alternatives have the required quality and can provide the same level of performance, their use may decrease the cost. However, their substitution must be approved in writing by the buyer beforehand.

2.4　Wadding

Wadding or filling is used to provide extra warmth to a garment. Filling usually refers to the fibres that are used to fill the space between the lining and the shell in a garment in an irregular state such as down filling and cotton filling. Quilting stitches are invariably used to constrain such filling fibres in position in the garment. Wadding usually refers to non-woven batts of fibre that may be cut to size and inserted between the shell fabric and the lining during garment making-up. As a batt, it is easier to handle and that's why it is widely used in clothing industries. The commonly-used types of wadding are polyester wadding and acrylic wadding, and they are usually specified by their weight in grams per metre squared. Wadding of $80g/m^2$ or above is typically used on winter clothing for very cold regions, and sometimes, lighter wadding is used for the sleeves. Some sheer non-woven sheets such as Lutrabond^{TM} could be used to reinforce the wadding and to prevent the fibres on the wadding from penetrating the lining or the shell fabric after being washed, causing unsightly pilling. Ordinary wadding of 40 g/m^2 may have low strength, and it is better to use Lutrabond^{TM} to reinforce it.

2.5　Other Accessories

Some other accessories may also be used in a garment. Shoulder pads can enhance the visual effect of a wider shoulder, which gives the wearer a "strong" image. Metal eyelets and rivets have been used to impart a masculine element, but they are now also used in ladies' garments.

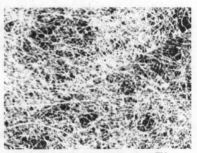

Fig. 3.3　Lutrabond^{TM}

Magnetic buttons are sometimes used on garments for easy fastening. Lace fabrics are used to enhance femininity, and are frequently used in blouses and ladies' underwear.

Drawstrings would be used on a hood, or at waist or hem to pull the hood or garment tight around the head or body respectively. The free end of the drawstring can be secured with either a double turn secured with lockstitch stitches, a heat seal, laser cut or a plastic sleeve (shoe lace end). A knot could also be used to secure the end of drawstring, but this is not recommended for children's garments. Cord ends could be used to stop the strings being drawn back into the string tunnel or channel and cord stops could be used to keep the string in a frapped state with the help of the spring in the cord stops (for spring loaded cord stops) or by the friction force between the string and the cord stops (for pig nose toggles, see Fig. 3.4).

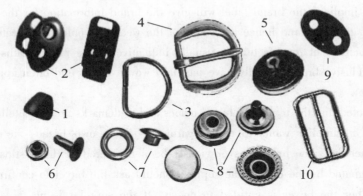

Fig. 3.4　Some commonly used accessories

1 — cord end　2 — cord stop　3 — D-ring　4 — buckle　5 — shank button
6 — rivet　7 — eyelet　8 — snap　9 — pig nose toggle　10 — slider

Elastic bands could be used at the waist or sleeve hem. Belts made of leather, artificial leather, or even the shell fabric could also be used, in which case buckles for the belt would be required. A belt can be retained with belt loops made of the shell fabric, or with D-rings, which are sometimes referred to as half rings.

No matter what accessories will be involved, clear specifications for them are necessary. The size, shape, colour and the materials used must be clarified and agreed during the business negotiation. Samples may be necessary to ensure that there is no misunderstanding about the type of accessory to be used. Similarly, for the colour, tone in tone and particularly where a special colour is designated, colour swatches or accessory samples in the specific colours may be essential, even for the "achromatic" accessories.

3　Labels and Hangtags

Attached to ready-made garments are labels and hangtags. These labels are sewn on the garments to provide the purchaser with some information regarding the garments, and hangtags are tagged onto the garments to confirm that the garments with hangtags have passed the final garment inspection.

3.1　Labels

There are many types of labels. These include the brand label, origin label, size label and washing label, most, if not all, of which should be found on a garment. Many garments only have a single label which gives information about the brand, origin, size and care instructions. Labels are located in different parts of the garment according to both the type of label and the type of garment. Many labels, other than the brand labels, are sewn into a seam within a garment as part of the seaming operation — a side seam is a typical example — this obviates the need for a separate label attaching operation.

3.1.1　Brand Label

As its name implies, the brand label will give the brand information for the garment. Since the "status" of a garment, and hence the status of the wearer, would be reflected by the brand information, its method of manufacture is usually indicative of the brand status. For example, expensive and exclusive brands usually have jacquard woven labels and often appear prominently within the garment.

It must be noted that the registered brand names and trademarks are the intellectual properties of the owner. If a supplier wants to export garments in his own brand, he should find out beforehand whether the same brand has previously been registered in the destination country. If the brand is designated by the buyer, the supplier should ask the buyer to confirm before signing the contract whether the buyer is entitled to do so. If the supplier is not able to obtain such information, he should insist on inserting a clause regarding the responsibility of the buyer on the property rights, saying, for example, "if the brand labels and hangtags designated by the Buyer should infringe the intellectual property right of a third party, or cause any disputes, the Buyer shall be fully responsible for the consequences and settlements".

3.1.2　Origin Label

The origin label is usually printed to show the country of origin of the garment. In international trade, to give the correct origin information is very important, especially when there is any constraint on garment trade. The Customs and Excise Office of the importing country can determine, from the origin information, whether duties should be levied at a general rate or a special rate.

In China, garments before exportation are legally required to undergo a mandatory inspection of the labels and hangtags by a competent government authority to prevent illegal entrepot trade.

If there is no information regarding the country of origin on garments and packaging, this would be referred to as the "neutral packing". Many countries such as the United States, the United Kingdom, Germany and Japan forbid the importation of neutrally packed goods and therefore, for garments exported to those countries, origin labels are essential.

3.1.3　Size Label

The size label is invariably printed. Clearly, the size information on the label is intended to enable the purchaser to choose the proper size of garment. For garments targeted at international travelers, so-called "international size labels" will be used which specifies the size used in the country where the market is located together with the corresponding equivalent sizes of other countries. Such a label greatly assists the traveller when choosing garments.

```
D36        D36
F38        F38
GB10       GB10
I42        I42
C8         C8
SP40       SP40
```

Fig. 3.5　"International size label"

3.1.4　Washing Label

Washing labels, or care labels, give the instructions on how to take care of the garments. The instructions are given in the form of some international textile care labeling codes as illustrated in Fig. 3.6. Generally, five symbols will be used, i. e. , a wash tub, a triangle, an iron, a circle and a square to indicate respectively the advice on the washing, bleaching, ironing, dry cleaning and drying process.

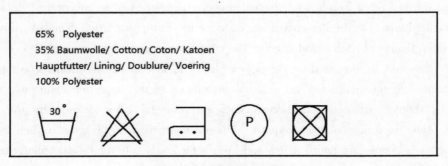

65% Polyester
35% Baumwolle/ Cotton/ Coton/ Katoen
Hauptfutter/ Lining/ Doublure/ Voering
100% Polyester

Fig. 3.6　Washing label

The care instructions are usually defined according to the composition of the shell fabrics; therefore many care labels would show the compositions of the shell and usually any lining as well. If there is a detachable lining, it may also need a separate care label since special instructions may be needed for that lining.

3.1.5　Other Labels

In addition to the labels giving the information on the brand, origin, size and care instructions, labels offering some other information may also be used.

a) **Composition Label.** Nowadays more and more people are concerned about the components used in the shell or lining. For the ordinary consumer, he or she will probably not be able to identify the components in a fabric from its appearance or the handle of the fabric. Therefore, unless such information has already been given on other labels, say the care label, a composition label should be included.

b) **Warning Label.** In some markets, garments such as nightwear must have a permanent label showing whether or not they meet the flammability standard and giving a warning "keep away from fire". For children's garments with small accessories, labels or hangtags showing a warning of a choking hazard may be required. Such warnings may appear on other labels, but sometimes a separate label or hangtag is essential.

c) **Ecological Label.** In many countries, especially developed countries, environmental issues are of increasing concern. Special regulations have been established and ecological labels are used for those garments that pass the accreditation on the ecology issues. There are many ecological labels, and the most famous one is probably the label of the OEKO-TEX standard. At present, ecological labels are not mandatory; however garments with ecological labels would no doubt be received favourably in the international market.

3.2 Hangtags

Most hangtags are made from artistically designed cards to present information to the customer about the garment. In addition to the information appearing on the labels, there is usually a bar code printed on them giving coded information on style number, batch number for the garment, etc.

If the hangtags are designated by the buyer, the information regarding the brand and origin should not infringe other people's intellectual property rights or violate government rules. Special attention must be paid to the designated bar code for the European Article Number, if any, since the first three figures of such a code denote the manufacturing country.

If the hangtags are designed by the exporter himself, the colours and designs on the hangtags should conform to the traditions and habitual preference of the importing countries; the literal instructions should satisfy the requirements of governmental rules defined by the importing countries. Most importing countries request the literal instructions to be given in their own official languages, for example, in both English and French for goods entering the Canadian market, and in Arabic for goods exported to the markets of Middle Eastern countries.

Both parties in a business negotiation should pay attention to the requirements for the labels and hangtags, that is, the requirements for the colours, designs, literal instructions, bar codes, materials, qualities, the positions where the labels or hangtags would be attached, etc. Whether such requirements are workable and how much these would cost should be taken into careful consideration.

Words and Phrases

shaped panel	成形的衣片
flat weft knitting machines	平型纬编针织机，横机
seamless ['siːmlɪs]	无缝的
fully-fashioned ['fulɪ-'fæʃənd]	全成形的
V-bed machine	横机（V 型针床机）
circular ['sɜːkjʊlə]weft knitting machine	纬编针织圆机
non-woven fabric	非织造织物
disposable [dɪs'pəuzəbl]garment	一次性服装
surgical ['sɜːdʒɪkəl]gown	手术袍
incontinent [ɪn'kɒntɪnənt]	（大小便）失禁的
drape [dreɪp]	悬垂性
leather ['leðə]	皮革
light weight fabric	薄型织物
silk habotai	真丝电力纺
heavy weight fabric	厚型织物
woolen Melton ['meltən]	粗纺毛麦尔登呢
natural ['nætʃərəl]fibre	天然纤维
man-made material	化纤材料
blend	混纺
textured ['tekstʃəd]	膨体的，变形的
luxurious [lʌg'zjʊərɪəs]	奢华的，豪华的
eye-catching	吸引眼球的，引人注目的
elegant ['elɪgənt] style	高雅的款式
apparel [ə'pærəl]	服装
non-irritating [nɒn-'ɪrɪteɪtɪŋ]	不会过敏的
absorbent [əb'sɔːbənt]	能吸湿的，能吸收的
polyester [ˌpɒlɪ'estə]	涤纶；聚酯
nylon ['naɪlən]	尼龙
elastomeric	弹性体的，有弹力的
Lycra ['laɪkrə]	莱卡（弹力纤维）
Spandex ['spændeks]	斯潘德克斯弹力纤维
synthetic [sɪn'θetɪk]material	合成纤维材料
porosity [pɔː'rɒsɪtɪ]	多孔性，透气性
extensibility [ɪkˌstensə'bɪlɪtɪ]	可延伸性
plain [pleɪn] weft [weft] knit	纬编平针
single jersey ['dʒɜːzɪ]	单面针织物
interlock [ˌɪntə'lɒk]	棉毛布（双罗纹）

double jersey	双面针织物
1×1 or 2×2 ribs	1+1 或 2+2 罗纹
snug [snʌg]	非常贴身的，舒适的
shape retention [rɪˈtenʃən]	保型性
shrinkage [ˈʃrɪŋkɪdʒ]	缩水(率)
curling [ˈkɜːlɪŋ]	卷边
plain woven structure	机织平纹织物
crepe de chine [kreɪpdəʃiːn]	双绉
taffeta [ˈtæfɪtə]	塔夫绸
cotton sheeting [ˈʃiːtɪŋ]	棉粗平布
cotton shirting [ˈʃɜːtɪŋ]	棉细平布
Tricot [ˈtrɪkəʊ] warp knitted fabric	特利考经编织物
stiffness [ˈstɪfnɪs]	硬挺性
abrasion-resistance [əˈbreɪʒən-rɪˈzɪstəns]	耐摩性
crease-resistance [kriːs-rɪˈzɪstəns]	抗皱性
poplin [ˈpɒplɪn]	府绸
chintz [tʃɪnts]	有光布，轧光布
palace [ˈpælɪs]	派力司
khaki [ˈkɑːkɪ]	咔叽
gabardine [ˈgæbədɪːn]/gaberdine	华达呢
serge [sɜːdʒ]	哔叽
twill [twɪl]	斜纹
yarn dyed	色织的;染色纱
floral [ˈflɔːrəl] pattern	花卉图案
propensity [prəˈpensɪtɪ] to pilling [ˈpɪlɪŋ]	起球的倾向
sewn-in lining	缝上去的夹里
polyester taffeta	涤丝纺
polyester pongee [pɒnˈdʒiː]	春亚纺
nylon taffeta	尼丝纺
rayon [ˈreɪɒn] satin [ˈsætɪn]	(人造丝)光缎羽纱
cotton flannel [ˈflænl]	棉绒布
viscose [ˈvɪskəʊs]	黏胶纤维
warp knitted meshes	经编网眼布
multiple continuous filaments	多孔长丝，复丝
locknit [lɒknɪt]	(经编)经平绒
reverse [rɪˈvɜːs] locknit	(经编)经绒平
buttoned-on	用纽扣脱卸的
weft knitted pile fabric	纬编割绒织物
down [daʊn] filling	羽绒芯

polyester wadding [ˈwɒdɪŋ] 　　　　　　　涤纶喷胶棉

trimmings 　　　　　　　　　　　　　　辅料，镶边料；修剪

metal-coated 　　　　　　　　　　　　　镀金属的

toxic [ˈtɒksɪk] 　　　　　　　　　　　　有毒的

sewing threads 　　　　　　　　　　　　缝纫线

joining 　　　　　　　　　　　　　　　缝合

edge neatening 　　　　　　　　　　　　拷边(收毛边)

topstitching [ˈtɒpstɪtʃɪŋ] 　　　　　　　缉缝

quilting [ˈkwɪltɪŋ] 　　　　　　　　　　绗缝

staple [ˈsteɪpl] spun [spʌn] 　　　　　　短纤维纺制的

core-spun [kɔː-spʌn] sewing thread 　　　包芯纺缝纫线

multifilament [ˌmʌltɪˈfɪləmənt] 　　　　　多孔丝，复丝

tenacity [tɪˈnæsɪtɪ] 　　　　　　　　　强度

elastic recovery 　　　　　　　　　　　回弹性

bulked continuous multifilament 　　　　膨体多孔长丝

textured continuous multifilament 　　　多孔变形丝，变形复丝

swimwear [swɪmweə] 　　　　　　　　　泳装

overedge [ˈəʊvəˌedʒ] stitching 　　　　　包缝

monofilament [ˈmɒnəʊˈfɪləmənt] thead 　单丝缝纫线

Nomex [ˈnəʊmeks] 　　　　　　　　　诺梅克斯(一种耐高温芳香族聚酰胺，杜邦公司品牌)

ticket number 　　　　　　　　　　　缝纫线标号

resultant [rɪˈzʌltənt] cotton count 　　　总(英)棉纱支数

metric [ˈmetrɪk] count 　　　　　　　　公制支数

tex [teks] 　　　　　　　　　　　　　特(号)数

denier [ˈdenjə] 　　　　　　　　　　　旦尼尔(旦数，纤度)

twist [twɪst] 　　　　　　　　　　　加捻，捻度

snarling [snɑːlɪŋ] 　　　　　　　　　缠结

knot [nɒt] 　　　　　　　　　　　　打节

dry clean 　　　　　　　　　　　　　干洗

overlock [ˈəʊvəlɒk] stitch 　　　　　　包缝线迹

raw [rɔː] edge 　　　　　　　　　　　毛边

fraying [freɪɪŋ] 　　　　　　　　　　散边

looper [ˈluːpə] thread 　　　　　　　　弯针线

tone in tone 　　　　　　　　　　　　配色

contrasting [kənˈtrɑːstɪŋ] colour 　　　镶色

colour fastness [ˈfɑːstnɪs] 　　　　　　色牢度

buttonhole [ˈbʌtnhəʊl] 　　　　　　　纽孔

embroidering [ɪmˈbrɒɪdərɪŋ] 　　　　　绣花

linear density	线密度
consumption [kən'sʌmpʃən]	耗用量
quotation [kwəʊ'teɪʃən]	报价
sample making	打样
front opening	前门襟开口
fastener ['fɑːsnə]	扣合件
style number	款号
article number	货号
leather-imitation button	仿皮扣
pearlescent [pɜː'lesnt]	珠光的
marble ['mɑːbl] grain [greɪn]	大理石纹理
shank [ʃæŋk] button	有脚扣
straight buttonhole	平头纽孔
straight buttonhole machine	平头锁眼机
eyelet ['aɪlɪt] buttonhole	圆头纽孔(凤凰眼)
eyelet buttonhole machine	圆头锁眼机
sub-contracted [sʌb-kən'træktɪd]	(工程)外包
wrapped-around [ræpt-ə'raʊnd] buttons	包扣
substitute ['sʌbstɪtjuːt] button	备用扣, 替代扣
spare [speə] button	备用扣
ligne [liːn]	(纽扣)号数
zipper teeth	拉链齿
zip slider	拉链头
zipper tape	拉链(底)带
slider body	拉链头座
pull-tab	拉柄
gripper	拉柄
pin lock	拉链头锁钩
pendant ['pendənt]	坠子
double-tag slider	双拉柄拉链头, 双柄拉头
metal zipper	金属拉链
zipper guard [gɑːd]	拉链挡布
nylon zipper	尼龙拉链
spiral ['spaɪərəl] zipper	尼龙拉链
plastic ['plæstɪk, 'plɑːstɪk] zipper	塑齿拉链
Vislon zipper	维士隆(塑齿)拉链
closed zipper	闭尾拉链
open zipper	开尾拉链
retaining [rɪ'teɪnɪŋ] box	针盒

insertion pin	插针
bottom stop	下止，尾掣
top stop	上止，头掣
metal claw [klɔ:]	金属爪扣
prong [prɒŋ]	戳销，爪扣
invisible zipper	隐形拉链
concealed [kən'si:ld]zipper	隐形拉链
snap [snæp]	揿纽，铐纽
press button	揿纽，铐纽
socket ['sɒkɪt]	揿纽阴扣
stud [stʌd]	揿纽阳扣
Velcro [vel'krəu]	尼龙搭扣
rivet ['rɪvɪt]	铆钉，工字扣
eyelet	气眼
substrate ['sʌbstreɪt]	底布
isotropic [aɪsəu'trɒpɪk]	各向同性
fusible ['fju:zəbl] interlining	热融衬
flat bed press	平板式压烫机
conveyorised [kən'veɪəraɪzd] press	传送带式连续压烫机
steam iron ['aɪən]	蒸汽熨斗
wadding ['wɒdɪŋ]	(纤维网状)填料
filling ['fɪlɪŋ]	(絮状)填料
batt [bæt]	纤维层，絮胎
acrylic [ə'krɪlɪk] wadding	腈纶喷胶棉
Lutrabond ['lu:trəbɒnd]	(卢特拉邦德品牌)纸膜
shoulder pad	肩衬，垫肩
magnetic [mæg'netɪk]button	磁性扣
plastic sleeve	塑料套管
shoe lace	鞋带
cord end	绳头
string tunnel ['tʌnl]	绳槽
string channel ['tʃænl]	绳槽
cord stop	绳塞，卡扣
frapped [fræpt]state	收紧的状态
spring loaded cord stop	弹簧型绳塞
pig nose toggle ['tɒgl]	椭圆型双眼绳塞
elastic band	宽紧带
artificial [ˌɑ:tɪ'fɪʃəl] leather	人造革
buckle ['bʌkl]	带扣，带夹

D-ring	半圆扣
half ring	半圆扣
slider ['slaɪdə]	(带子上的)滑扣
colour swatch [swɒtʃ]	色样，染色样布
achromatic [ˌækrəʊ'mætɪk]	非彩色的
label ['leɪbl]	标签
hangtag ['hæŋtæg]	吊牌
brand [brænd] label	品牌标签
origin ['ɒrɪdʒɪn] label	原产地标签
size label	尺码标签
washing label	洗涤标签(洗水唛)
jacquard [dʒə'kɑːd]	提花的
trademark ['treɪdmɑːk]	商标
intellectual [ˌɪntə'lektjʊəl]	
property ['prɒpətɪ] (right)	知识产权
general rate	普通税率
special rate	特别税率
mandatory ['mændətərɪ] inspection	强制检验
competent government authority	政府主管部门
illegal [ɪ'liːgəl] entrepot ['ɒntrəpəʊ] trade	非法转口贸易
neutral ['njuːtrəl] packing	中性包装
international size label	国际尺码标签
care label	(洗涤)保养标签(洗水唛)
composition label	成分标签
warning ['wɔːnɪŋ] label	警示标签
nightwear ['naɪtweə]	睡衣
flammability [ˌflæmə'bɪlətɪ]	易燃，可燃性
choking ['tʃəʊkɪŋ] hazard ['hæzəd]	窒息危险
ecological [ˌekə'lɒdʒɪkəl] label	生态标签
accreditation [əˌkredɪ'teɪʃən]	认证
OEKO-TEX standard	OEKO-TEX 标准
bar code	条形码
European [ˌjʊərə'piː(ː)ən] Article Number	欧洲物品编码
literal ['lɪtərəl] instructions	文字说明
official language	官方语言

Exercises

A. Please use "T" or "F" to indicate whether the following statements are TRUE or FALSE：

(1) The looper threads within edge neatening stitches are often textured multifilament yarns which provide the strength, extensibility and the softness required.　　　　　　(　　)

(2) A thread used for joining fabrics should have sufficient strength, and should not shrink more than that of the garment pieces to be joined.　　　　　　(　　)

(3) The zipper which has teeth formed with a spiral of nylon monofilament coils is referred to as a "plastic zipper".　　　　　　(　　)

(4) Since the grain lines of knitted or woven interlinings must follow those on the garment pieces, relatively lower fabric utilisation would be inevitable.　　　　　　(　　)

(5) The length of zippers for coat front plackets must be varied according to the length of the garments.　　　　　　(　　)

(6) The main purpose of using cord stops is to stop the strings being drawn back into the string tunnel.　　　　　　(　　)

(7) If there is information regarding a country on either garments or packaging, this would not be referred to as "neutral packing".　　　　　　(　　)

(8) An origin label must show the country where the garment is finally processed.　　(　　)

(9) One of the functions for quilting is to hold the wadding or filling in a garment in position.
　　　　　　(　　)

(10) If a belt is involved in certain style of garment, belt loops must be used to hold the belt.
　　　　　　(　　)

B. Please select the word(s) or phrase(s) that makes the statement correct：

(1) The disadvantages of knitted fabrics for garment making are _____.　　(　　)
　a) relatively poor shape retention
　b) relative lower extensibility
　c) relatively higher shrinkage
　d) edge or end curling

(2) "Pilling" on sweaters refers to a phenomenon in which _____.　　(　　)
　a) a small piles of dirt accumulate on the fabric due to static electricity
　b) small balls of fibres on the fabric surface are caused by the entanglement of fibres
　c) loops of yarn are pulled out and protrude from the fabric

(3) _____ is a commonly used twill structure for outerwear.　　(　　)
　a) Chintz　　　b) Gabardine　　　c) Khaki　　　d) Poplin

(4) _____ is one of the main functions of the threads in clothing production.　　(　　)
　a) Fusing　　b) Joining　　c) Edge neatening　　d) Quilting

(5) A button with a size of 32 ligne has a diameter of _____.　　(　　)
　a) 4/5ths of an inch　　　　b) 32.00 mm

   c) 5/8ths of an inch   d) about 20. 32 mm

(6)Which of the following accessories can be used as fasteners?   (   )

   a) Shank button   b) Plastic zipper

   c) Velcro   d) Eyelet

Reading Materials

More Warmth When the Days Turn Colder

A. Foul-weather Quarter-zip Pullover

Windproof, Water repelling, and warm, our quarter-zip pullover offers superior protection from the elements. Soft, faux-suede trim inside the neck, on the shoulder patches, and at the pockets. Sizes S–XXL* Shell: 100% wool. Lining: 100% polyester. Washable.

B. Fairbanks Houndstooth Shirt

These thick cotton shirts are heathered for a handsome look. Washable faux-suede trim. Sizes M–XXL* 100% cotton. Washable.

C. Ultimate Foul Weather Sweater

A shooting sweater designed to deal with the elements on windy fall days. This windproof and water-resistant shooting sweater has a lined body and lined collar to block the wind and keep you comfortable and warm while concentrating on the targets. Cotton twill gun patches cover both shoulders, while ribbed knit at the cuffs and bottom hold out drafts. Body is lined in plaid with stretch for comfort and Teflon® coated wool for water resistance. Handwarmer pockets on lower body; interior zipped security pocket. Sizes S–XXL*. 100% wool. Washable.

· Soft touch, collar & handwarmer pockets are lined with fleece.

· Laugh at the wind, ingenious liner stops gusts in their tracks, but still breathes.

· Stay strong, rib-knit cuffs & hem lock in warmth & won't stretch out over time.

· New! Interior zip pocket, for extra security.

*S(34–36), M(38–40), L(42–44), XL(46–48), XXL(50–52)

A.  B.  C.

—excerpted from *Orvis UK* 2022 *Late Winter Men's Catalogue*, page 12 and 18.

【参考提示】

1. 以上摘自 ORVIS 的英国公司(orvis. co. uk)发布的冬季男装目录。Orvis 公司是以其创始人之姓命名的, 总部在美国佛蒙特州。服装是该公司经营的主要商品之一。

2. quarter-zip "四分之一拉链",指"开约四分之一衣长并装拉链的门襟"。

3. the elements 恶劣天气。这个表达中 element 必须是复数并使用定冠词。

4. faux-suede 仿麂皮。

5. shoulder patch 肩部挡布。

6. Fairbanks houndstooth shirt 可译为"Fairbanks 千鸟格衬衫"。houndstooth 指"犬牙花纹(千鸟格)",fairbanks 是人名或地名。

7. heathered 被做成"麻点夹花"、杂色效应。

8. shooting sweater 意思是 sweater for shooting,指"射击运动时穿的针织衫"。

9. ribbed knit 和 rib-knit 都是指"针织罗纹"。

10. draft 穿风(传统英式英语中用 draught)。

11. plaid 方格布,方格呢。

12. handwarmer pocket 暖手衣袋。

13. ingenious liner stops gusts in their tracks, but still breathes 该句意思指"衬里巧妙地阻挡住穿过针织面料的风,但仍能透气"。其中 liner 指衬里,gusts 原指"阵风"。从该句前的 Laugh at the wind 足以看到该文撰稿人的风趣。

14. ...won't stretch out over time 指(罗纹针织袖口和下摆)"久拉却不松弛"

15. S 即 Small;M 即 Medium;L 即 Large;XL 即 Extra Large;2XL 或 XXL 即 Double Extra Large。它们指的都是服装尺码。另外还有尺码 XS 即 Extra Small 及 3XL 或 XXXL 即 Triple Extra Large。

Focus on Fabrics

Feast your eyes on the latest fabrics for this season's sewing!

Flannel for Apparel

The softest, coziest flannel is now available from AGF! Made with the highest standards, this new flannel is 100% double-brushed cotton. Both sides are super soft and ideal to keep you warm in the chillier months. You can also feel comfortable that it's OEKO-TEX certified so it's produced without compromising the planet. Enjoy using the quality fabric for making quilts, garments like pyjamas and other craft projects.

—excerpted from *Modern Sewing starts here*, edition 18-October 2021, page 9.

【参考提示】

1. flannel 如果是毛料的,则通常称为"法兰绒";如果是棉料的,则一般译作"棉绒布"。从文中介绍的材料可知,这里涉及的应该是"棉绒布"。

2. AGF 即美国知名的面料商 Art Gallery Fabrics。

3. 100% double-brushed cotton 全棉且双面刷绒。

4. OEKO-TEX certified 指通过国际纺织及皮革生态学研究和检测协会(the International Association for Research and Testing in the Field of Textile and Leather Ecology)的 OEKO-TEX 生态标准认证的。

Sewing Workshop Collection "Marceau Tee"

Loose-fitting, hip-length top with shifted right and left sides, ready-to-wear neck binding, two-piece sleeves with cuffs and pleated detail on upper sleeve. Designed for busts 31″–46″.

Recommended Fabrics：Light to medium weight knits— 2-way and 4-way stretch.

Recommended Notions：Thread.

Yardage requirements, 1 fabric：

54/60″: 1. 75 yards for sizes XS–M

          2. 25 yards for sizes L–XXL

Yardage requirements, 2 fabrics：

54/60″: 7/8 yards each fabric for XS–M

          1. 385 yards each fabric for L–XXL

You can run the neckline banding either direction—with the stripe or perpendicular to the stripe.

—excerpted from *Vogue Fabrics*, Winter/Holiday 2021, page 17.

【参考提示】

1. *Vogue Fabrics* 是位于美国芝加哥的美国主要的时尚面料零售及批发商 Vogue Fabrics Store 线上定期发布的商品目录，介绍他们所经营的服装面料、辅料以及推荐的相关服装样板款式。(http://www. voguefabricsstore. com)

2. Sewing Workshop 是 Vogue Fabrics Store 公司旗下众多独立样板设计团队之一。上文介绍的是来自 Sewing Workshop 的系列为 Marceau Tee 的作品。

3. shifted right and left sides 左右两侧错位(设计)。

4. light to medium weight knits 薄型至中等厚度的针织布。

5. 2-way and 4-way stretch 这里的 "2-way stretch"指"纵向或横向弹性"(纬编针织物通常具横向弹性)，"4-way stretch"指 "纵向和横向都具有弹性"。严格来说，这里的 and 应该换成 or。

6. Notion 在美式英语中有时作"小物件(针、线、纽扣、绲边带等)"解释，这里应该表示如果购买这款服装样板，那么商家推荐的辅料为它家的缝线。

7. Yardage 用料码长。该款服装作为左右不对称设计，更可以考虑左右使用不同颜色的条纹布，因此文中的"Yardage requirements, 2 Fabrics"就是这种情况下的用料码长。

8. 54/60″ 这里应该指以英寸计的用料门幅。不过这里的 6 英寸(约 15 厘米)容差似乎太大。

9. You can run the neckline banding either direction－with the stripe or perpendicular to the stripe 指的是 "你可以采用领圈料走向为顺(布料)条纹或垂直(布料)条纹的"。

10. """为英寸("'"为英尺)。国外常用码(yard)来计量布料长度，用英寸计量门幅宽度。1 码等于 0.9144 米，1 英寸等于 2.54 厘米。

Alter a Men's Shirt: Take in the Back
Three fixes to give your off-the-rack shirt a sharper fit
By Vanessa Nirode

For many men, finding a button-front dress shirt that fits correctly can be difficult. If a shirt fits at the collar, the sleeves may not be the correct length, the shoulders may be too wide, or the body too full. "Standard" sizing for men's apparel has some of the same problems as women's sizing: The proportions are often ideal for only a small portion of the population.

Although it's possible to hide some fit problems under a suit jacket, casual office and leisure settings rarely call for jackets these days. To look neat, pulled together, and modern, it's worthwhile to assess whether your shirts have enough ease where it's needed, but not too much, and that the body and sleeve lengths suit your build.

Many brands sell dress shirts with descriptive labels such as "regular" or "traditional" fit, "slim" fit, and "athletic" fit. (Not all brands offer each of these cuts.) The regular fit is usually a fuller cut with pleats in the back below the yoke. A traditional Brooks Brothers shirt is cut this way. Slim-fit shirts typically do not have these pleats. They usually have side-back darks to create a closer silhouette at the back waist. John Varvatos brand shirts are a good example of this kind of fit. The athletic fit has a fuller cut than the regular, with wider sleeves to accommodate muscular arms. Even with these options, a well-fitting shirt off the rack can prove to be elusive. But, with a little altering, most fit issues can be resolved by someone with intermediate sewing skills.

—excerpted from *Threads* #216, Winter 2021, page 40.

【参考提示】

1. off-the-rack 工厂批量生产的，非度身定制的。

2. John Varvatos 著名男装设计师，国内有的译为"约翰·瓦维托斯"。上文的 Brooks Brothers 也是有名的男装品牌，国内也有的译为"布克兄弟"。

CHAPTER *4* WORKING SHEETS

1 The Function of Working Sheets

Working sheets are sometimes called specification sheets (SPEC sheets). Generally speaking, each garment style should have one working sheet. The function of working sheets is to provide the manufacturing department of the clothing factory with the detailed instructions for making-up each garment. The material descriptions on the working sheets could also give suppliers information regarding the specifications and consumption for the shell fabrics, linings and accessories so that the cost of the particular garment can be estimated. Working sheets are usually supplied by the party who has designed the garment. In garment trading, the working sheets would be provided by the buyer if the transaction with the supplier is agreed on the basis of buyer's samples. In such a case, the garment supplier should read the working sheet carefully since the specifications and instructions in the working sheets form part of the contract. The technical staff in the clothing factory would also produce their " own" working sheets in their own language according to the specifications from the buyer's working sheets and the technical parameters resulting from the sample making.

The layout of the working sheets will, invariably, differ from company to company. Some working sheets may include detailed information whereas others may be very brief. For example, if the transaction concluded involves samples, and if both parties concerned are familiar with each other, only a brief working sheet may be necessary assuming that the supplier understands the exact requirements. However, it is still advisable for detailed information to be provided if the style of the garment is complex.

2 Contents of a Working Sheet

A working sheet should clearly specify the style number, to which the working sheet relates. If the pattern number differs from the style number, the pattern number should also be shown on the working sheet. This may happen when one set of basic patterns generates several garment styles that differ because of the materials used to make them. There is no hard and fast rule as to what must be specified in a working sheet, but it is advisable that the information given in a working sheet should be clear enough so that the garment could be produced correctly in accordance with the instructions. A working sheet providing a comprehensive set of instructions would include the following details, so that there should be no misunderstanding about how the garment should be made:

a) material specifications and consumption;

b) garment working sketches;

c) technical details;

d) a size chart, if it is not supplied separately;

e) a list of the patterns required, if they are to be supplied by the buyer;

f) other information as necessary, e. g. the label specifications and their locations within the garment, packing instructions and marking requirements for packages. Such information may also be supplied elsewhere.

2.1　Material Specifications and Consumption

The material specifications and consumption defined in the working sheet should cover all the materials required to produce the garment. It should include details for shell, lining and interlining fabrics, sewing threads, buttons and other fastenings and accessories such as belts. The locations for attaching the fastenings and other accessories should also be defined.

a) **Material specifications.** The types of material shown in a working sheet are usually specified under a general name with the detailed specifications about the materials provided in the form of samples or in other documents or agreements. For example, without a sample, a button specification of "ST332" is likely to be meaningless. Furthermore, if there are no other documents or agreements specifying yarn counts and fabric area density, etc., a specification on the working sheet of "polyester pongee" would not be sufficient to define the fabric to be used.

b) **Material colours.** In garment production, it is more convenient to indicate a colour by a number than its name. Of course, all colours involved must be specified with reference to a colour card or colour swatches. The colour of the shell fabric (garment colour) is usually given in the garment order, but if contrasting colours are involved, how the colours should be matched must be specified through a colour-matching table or by numbering the garment colour as a combination of the numbers of the matched colours. For example, if the primary colour of the shell fabric is Col. 2 and the (secondary) contrasting shell colour matching primary Col. 2 is Col. 5, the garment colour number would be Col. 25.

If the colour of the lining or an accessory is specified on the working sheet as "t. i. t. " (an abbreviation for the phrase "tone in tone"), it means that the colour of that material must comply with the colour of the shell where the material is located. Under such circumstances, care must be taken, especially when the garment has contrasting colours. If the colour of the lining or an accessory is defined as a specific colour number, it means that this particular colour should be used for that material irrespective of the variation in the colours of the shell fabric(s).

c) **Material consumption.** The material consumption specified in a working sheet relates to a particular size, usually a medium size. If the range of sizes in an order is evenly distributed, the average material consumption per garment for that order generally approaches the consumption for the medium size garment. If a working sheet with accurate information regarding the material consumption could be available in the business negotiation, it would be helpful to both parties. However not all working sheets would necessarily contain the information regarding the material consumption at this stage of the negotiation. If such information is unavailable, the material consumption would either have to be determined through the sample making, or estimated by experience.

2.2 Working Sketch

There may be one or more working sketches (working drawings) in a working sheet. The working sketches depict the appearance of the front and the back of the garment. For garment tops, if the inside of garment bodices need to be shown, the shoulder seams would have to be shown opened to give a clearer view (see the working sketches in the working sheet attached at the end of this chapter). Some brief details could be included on the sketches, such as those components in contrasting colours and the position and width of topstitching.

2.3 Technical Details

Technical details provide instructions to make-up a particular garment style. The details should cover:

a) precise locations for each component within the garment;

b) accurate and comprehensive instructions for each sub-assembly and final assembly operations;

c) specifications for stitch types, seam types, stitch densities, seam allowances, instructions for seam neatening and topstitching, etc.

If these instructions regarding the garment making-up can be determined from garment samples or other documents, or according to the customary practice agreed in previous contracts with the same buyer/manufacturer, then the technical instructions shown in the working sheet may only need to be minimal. For example, some requirements, say stitch types, stitch densities, seam types and seam allowances, may have been specified in other documents or may be determined from the samples, in which case, the working sheet may not need to provide such information. However, if neither samples nor documents (including the working sheet) provide information about the technical requirements for the specific garment, it is advisable that both parties, the buyer and the supplier, should clarify and agree them preferably in the contract negotiations and certainly before production commences, even though the manufacturer/supplier may simply follow their customary practice used in garment production.

2.3.1 Stitch Types

Although there are various types of stitches, they are all formed by interlacing, interlooping or intralooping the sewing threads – this forms the basis for the classification of stitch types.

When interlacing, a loop formed by one thread passes over or around another loop formed by a different thread; when interlooping, a loop formed by one thread passes through another loop formed by a different thread; when intralooping, a loop formed by one thread passes through another loop formed by the same thread (see Fig. 4.1).

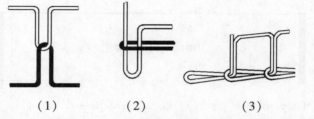

（1）　　　　　（2）　　　　　（3）

Fig. 4. 1　Interlacing　（2）Interlooping　（3）Intralooping

Stitches are divided into six classes, and each class contains several types of stitches. The characteristics of these six stitch classes are presented in the following table.

Table 4. 1　Characteristics of the six stitch classes

Stitch Class No.	Name	Characteristics	Example
100	Chain stitch	A loop or loops of thread or threads are passed through the fabric（s）and secured by intralooping with the succeeding loop or loops after they are passed through the fabric(s) to form a stitch.	
200	Hand stitch	One thread is passed through the fabric（s）from one side to the other and vice versa with the appropriate stitch length to form a stitch.	
300	Lock stitch	Loops of one（or a group of）thread（s）are passed through the fabrics where they are secured by interlacing with another thread or threads of another group to form a stitch.	
400	Multi-thread chain stitch	Loops of one group of threads are passed through the fabrics and are secured by interlacing and interlooping with loops of another group to form a stitch.	

(Continued)

Stitch Class No.	Name	Characteristics	Example
500	Overlock stitch (overedge or edge seaming)	Loops of one group of threads are passed through the fabric(s), secured by intralooping before succeeding loops are passed through the fabric(s), or secured by interlooping with loops of one or more interlooped groups of threads before succeeding threads of the first group are again passed through the fabric(s). Loops from at least one group of threads pass around the edge of the fabric(s).	
600	Flat seam stitch	Loops of one group of threads are passed through loops of the second group already cast on the surface of the fabrics and then, passed through the fabrics where they are interlooped with loops of a third group of threads on the underside of the fabrics.	

Stitches within the Class 100 have excellent extensibility and a neat appearance but will unravel easily if the thread is broken. One of the commonly-used types is stitch Type 101, which is a single thread chain stitch, and is extensively used for basting, i. e. sewing with temporary stitches. Stitch Type 103 (blind stitch) is not visible from one side of the garment and is extensively used for hemming (See Fig. 4. 2).

Stitches within Class 200 have little application in mass production. In order to achieve the "hand stitch" effect on men's or women's tailored garments, some sewing machine companies have developed hand-stitch machines which can produce the stitches under Class 200 with a double-pointed needle. The most common stitch type in this class is stitch Type 209 (See also Fig. 4. 2).

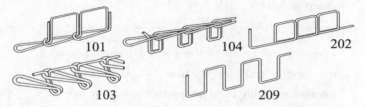

101 104 202

103 209

Fig. 4. 2 Examples of stitch types (Class 100 and Class 200)

In stitch Class 300, the most widely used is stitch Type 301, commonly referred to as lockstitch, which does not easily unravel but usually has limited extensibility. Its extensibility is affected by the balance of the stitch. Ideally, the interlacing should occur between the fabrics to be joined. However, if the interlacing occurs either on top or below the fabrics, the extensibility is restricted principally to the extensibility of the thread that lies straight (see Fig. 4. 3). The balance of the stitch is determined and controlled by the tensions on the respective threads. The fact that it can be started and finished at any point on the fabric surface makes it very suitable for sewing darts and pleats, or attaching patch pockets, etc. However, at both ends of the seam formed by stitches of stitch Type 301, back tacking (sewing backwards and forwards for a few stitches) is often used to secure the seam. Stitch Types 304 and 308, widely used in foundation garments and in swimwear, are zigzag stitches, and the zigzag configuration gives a significantly higher extensibility than stitch Type 301 (See Fig. 4. 4).

a　　　　　　　　　　　　b

Fig. 4. 3　Thread interlacing

a) correct interlacing　　　b) incorrect interlacing

　　Among the stitches of Class 400, stitch Type 401, a double-thread chain stitch (double locked chain stitch), is the second most widely used stitch type in clothing production. It is commonly used to join the centre seat seams of trousers, or attach waistbands, etc., due to its higher seam strength and extensibility. Stitch Type 404 is a zigzag stitch and has a similar structure to stitch Type 401. Stitch Type 406 is a multi-thread chain stitch and it is often referred to as a cover stitch. Class 400 will not unravel as easily as Class 100 (See Fig. 4. 4).

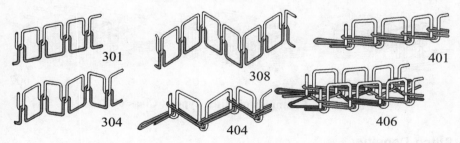

Fig. 4. 4　Examples of stitch types (Class 300 and Class 400)

　　Stitches in Class 500, referred to as overlock stitches, are widely used in garment production to bind the trimmed edges of a fabric and for joining stretchy garment pieces. Stitch Types 502 and 503 are two-thread overlock stitches. Stitch Types 504 and 505 are three-thread overlock stitches which are very popular stitch types. Stitch Types 512 and 514 are double-needle four-thread overlock stitches, and stitch Type 516 is a safety stitch, which in fact is a combination of stitch Types 401 and 505 (See Fig. 4.5).

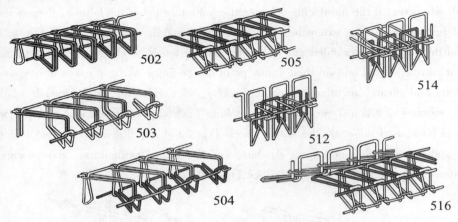

Fig. 4. 5   Examples of stitch types（Class 500）

Stitches of Class 600 are flat seam stitches, which provide good extensibility necessary in seaming knitted fabrics together with a flat and comfortable seam appearance and good seam strength. Certain Class 600 stitches are referred to as cover stitches, but it should be noted that, unlike cover stitches in Class 400, they cover both the top and bottom of the seam. The commonly used types are stitch Types 602, 605 and 607. Most stitches under Class 600 are formed with three groups of threads. One group bridges the butted join on the face of the fabrics, and the second group bridges the butted join on the reverse of the fabrics; the two groups are interconnected through the fabrics by the third group, i. e. the needle threads formed by needles. Two, three and four needles are involved respectively in stitch Types 602, 605 and 607 (See Fig. 4. 6).

Fig. 4. 6   Examples of stitch types（Class 600）

2.3.2 Stitch Density

Stitch density is the number of stitches per unit length of seam. It is determined according to the specific requirement for garment making-up to provide adequate seam strength, extensibility and appearance. The stitch density may be decided from experience or during the manufacture of a sample garment. It is dependent on the amount of the fabric(s) that are moved forward between stitch cycles—the greater the movement, the lower the stitch density. The stitch density also affects the speed of the operation and thread consumption and hence, the production cost. Too low a stitch density would also affect the quality of a garment. Different types of stitches, different seaming operations and different fabrics require different stitch densities.

2.3.3　Seam Types

The main function of a seam in garment production is to join the fabric components together. A desirable seam should have adequate strength, so that the joined garment pieces will not be torn apart by the stresses imposed when the garment is worn, and it should have a neat appearance without puckering, which is one of the common sewing faults in garment production (see Section 3.2.2 of Chapter 6), or other apparent sewing faults. Furthermore the seam must be at least as extensible as either the fabric(s) or as required by the movement of the related part of the garment.

Seams are divided into eight classes, within which there are many variations. In clothing manufacture, the following four main seam types are commonly used:

a) Superimposed seam (Class 1);

b) Lapped seam (Class 2);

c) Bound seam (Class 3);

d) Butted or flat seam (Class 4).

Fig. 4.7 illustrates the four main seam types and their variations. Seam types within Classes 5 to 8 could also be used in clothing production and Fig. 4.8 shows some examples. Please note that the dotted lines in these figures denote the direction of the main part of each fabric as opposed to the cut edges.

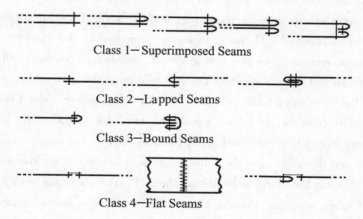

Class 1—Superimposed Seams

Class 2—Lapped Seams

Class 3—Bound Seams

Class 4—Flat Seams

Fig. 4.7　Examples of seam types under Class 1 to 4

Seams may be defined by a five-digit number. The first digit denotes the seam class, the second and third digits define the different fabric configurations and the fourth and fifth digits denote the different needle penetrations and/or mirror images in the configuration of the materials used in the seam. So, for example, the simplest superimposed seam would be defined as 1-01-01 (see the first diagram under the heading "Class 1-Superimposed Seams" of Fig. 4.7).

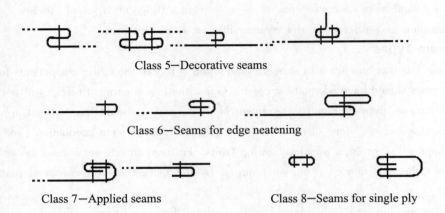

Class 5—Decorative seams

Class 6—Seams for edge neatening

Class 7—Applied seams   Class 8—Seams for single ply

Fig. 4. 8 Examples of seam types under Class 5 to 8

2.3.4 Edge Binding and Seam Neatening

Except for fashioned knitted garments, in which the garment components are shaped by the technology of narrowing and widening during knitting, a garment component in a cut-fashioned garment will have raw edges after it is cut. The raw edges may need to be bound or oversewn (neatened) to prevent yarns from slipping out from the raw edges, or the fabric will unravel along the raw edges, which may destroy the seams. For a lined garment, seam neatening is usually not necessary because all the raw edges of the garment components are enclosed and protected between the shell and the lining. For unlined garments, seam neatening is necessary.

There are several methods of seam neatening, and the method used depends on the type of fabric, the seam types involved and the style and construction of the garment. Overlocking is widely used for seam neatening as the raw edges of garment pieces are well bound by the overlocking stitches without unduly affecting the extensibility of the garment pieces. Binding can be used to neaten the opening or edges of a garment. This uses seams under Class 3, whereby a strip of fabric, usually bias-cut, is folded around the raw edge of the fabric to be bound and stitched through both fabrics to secure the binding. One example is the use of a rouleau to bind the neckline of a dress and the other example is the use of bias binding to enclose all raw edges of a detachable lining. Cutting the fabric with pinking shears to create a zigzag cut edge (pinking) is widely used to cut fabric samples. Pinking can also be used for neatening seams of non-fraying fabrics. There are some seam types, in which the raw edges of the garment pieces will be double-turned and stitched (6-03-01 for example. See the second diagram under the heading " Class 6-Seams for edge neatening" of Fig. 4. 8). Thus the raw edges are kept inside the seam. This is sometimes referred to as a "clean finish". If the raw edges of two garment components are turned and embedded in the seam, this is referred to as "self-neatening", of which the French seam (1-06-01, the last diagram under the heading "Class 1-Superimposed Seams" of Fig. 4. 7) is a typical example.

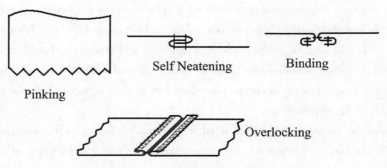

Fig. 4.9　Seam Neatening

2.3.5　Seam Allowance

Seam allowance is essential for a seam. This is defined as the additional amount of fabric required at the edge or within a pattern to enable the fabric components to be joined together. It is, in effect, the distance from the line of stitches to the cut edges of the fabric. Insufficient seam allowance may mean that the line of stitches will be pulled out from the seam when it is stressed whilst being worn. Too much seam allowance wastes fabric. Consequently, the seam allowances adopted affect the fabric consumption and the seam strength. If the patterns for the garment sample are supplied by the buyer, the seam allowances will be indicated by notches on the patterns. If the paper patterns are to be cut by the clothing factory, the seam allowances should be agreed during the contract negotiations and subsequently specified in the working sheet.

The seam allowance depends on the type and the structure of fabrics to be joined, the shape of the seam lines and the type of the seam itself. Loosely woven or knitted structures may need relatively bigger seam allowances than tighter fabrics, and a curved seam line would usually require a narrower seam allowance than a straight seam line.

2.4　Size Chart

The size chart attached to the working sheet lists the measurements for the distances between key garment locations over the range of sizes for the specific garment. These are related to the anthropometric data for the population that the garment is intended to fit. The measurement locations in the chart constitute criteria by which the finished garment is inspected to ensure that the garments are manufactured to the correct dimensions for their intended size.

It is very important that the chart should specify clearly the measurement requirements, or otherwise, disputes may arise due to misunderstandings regarding measurement locations. To illustrate this, consider the "across back" measurement. The measurements made at either higher or lower positions across the back are likely to record different lengths. Therefore, some companies would specify clearly in the size chart that the "across back" measurement is to be made "4 inches down from the centre back neck". Other companies may provide diagrams, in which the measurement locations are clearly marked.

It is also very important that the chart should specify clearly the measurement tolerances as well. For many manufacturers and particularly clothing manufacturers to produce garments to precise length and width measurements is unfeasible because of the nature of the materials and the manufacturing process. Suppose the bust girth for a certain size of garment is specified as 120 cm, the final bust girth of the finished garment is unlikely to be exactly 120 cm. Tolerances specify the limits to the acceptable range within which a specific measurement should fall and therefore become the criterion about whether the garment under inspection should be accepted or rejected. The values for the tolerances are a compromise between what is achievable in the mass production and the accuracy of fit required.

The number of measurements that need to be made during both intermediate and final inspections will vary and is particularly dependent on the accuracy of fit required for the garment. loose-fitting garments invariably require fewer measurements than tailored or tight fitting garments. Another reason may be that in the case where paper patterns for all sizes are supplied by the buyer and if counter samples have been approved by the buyer, simply by checking only a few important measurements such as the centre back length, sleeve length, bust girth and waist girth, the inspector from the buyer could ascertain whether the measurements of the garment are within the tolerances and can therefore satisfy the requirements defined in the contract. However, if the garment supplier is requested to grade the patterns and if the buyer cannot supply a grading chart, the size chart should contain measurements for a sufficient number of locations to ensure that the supplier can deduce the grading rules needed to produce the garments to the correct sizes and dimensions specified in the order.

3 Abbreviations and Simplified Words in the Working Sheets

The use of easily understood abbreviations and simplified words when drafting a working sheet is often preferred to save time. Universally accepted abbreviations are usually readily understood but simplified words are sometimes confusing depending on their derivation. The following rules to simplify words are commonly followed when working sheets are drafted:

a) Use the first letters of the words in a phrase or use the first letters of the component parts in a compound word, e. g. :

CB→Centre Back;   CF→Centre Front;   CBF→Centre Back Fold;
CFF→Centre Front Fold; RS→Right Side;   AH→armhole;
BP→bust point;   SS→shoulder slope; ZO→zip out;
HSP→higher shoulder point

b) Retain the first syllable or the first and second syllables, or retain the first syllable or the first and second syllables and keep the first consonant of the following syllable, e. g. :

COL→colour;   DIST→distance;   Consump. →consumption

c) Retain the first and the last letters of a short word, e. g. :

NK→neck;   LT→light;   DK→dark;
DN→down;   BK→back;   FM→from;

d) Omit all vowels except the vowel at the beginning of a word and retain all or essential consonants, e. g. :

HV→have; SLV→sleeve;　ACPT→accept;　　QNTY→quantity;

PLKT→placket;　　　　FRT→front;　　　DBL→double

e) Use a letter or letters to stand for a word with similar pronunciations, e. g. :

N→and;　　　　　　　B→be;　　　　OZWZ→otherwise

f) Use simplified suffixes, e. g. :

G→-ing;　　　　　　　D→-ed;　　　　BL→-able, -ible;

T, MT→-ment;　　　　N, TN→-tion

The following size chart and working sheet are typical examples of those used in clothing manufacture. The company is fictitious.

John & Grace Fashion Company

1234 Victoria Street, Hong Kong

SIZE SPECIFICATIONS

STYLE NO.: 20345

DESCRIPTION: Pants

UNITS: inch

MEASUREMENT	SIZES			
LOCATIONS	S	M	L	XL
Waist relaxed	28	30	33	37
Waist extended	42	44	47	50
Low hip (8" below band)	42	44	47	50
Waistband width	1/2	1/2	1/2	1/2
Front rise (below band)	11 1/4	11 3/4	12 1/4	12 3/4
Back rise (below band)	14 1/2	15	15 1/2	16
Inseam	29 3/8	29 1/2	29 5/8	29 3/4
Outseam	38 1/2	39	39 1/2	40
Thigh (1" below crotch)	26 1/4	27 1/2	29 1/4	31 1/2
Leg opening	17 1/2	18	18 1/2	19
Hem height	1	1	1	1
Hanger loop length	6	6	6	6
Hanger loop width	1/4	1/4	1/4	1/4

* **Remark**: The tolerances are the same as those used in orders for Winter 2021.

John & Grace Fashion Company

1234 Victoria Street, Hong Kong

WORKING SHEET

Season:Winter 2021

Style No.:2564   **Size**: 8

Description:Ladies' jacket with buttoned-on detachable lining   **Length**:80cm

Material   **Consumption**:

Shell:	**Width(cm)**	**consump.（cm）**
T/C Poplin, Art. 2520		
I	150	215
II	150	80
Lining:		
Polyester Taffeta, T190		
Plain, t. i. t.	150	120
Quilted, on 40 gr.	150	60
Wadd. +Lutrabond		
7 cm long stitch, t. i. t.		
Wadding:		
40 gr. Wadd.	150	50
Interlining:		
Art. 2300	100	85

Buttons:				
Art. No.	QNTY	Size	Colour	Position
ST234	7	32L	t. i. t.	
+ 1 substitute button				
ST122	17	24L	achrom.	Lining

Threads：

	Col.	Count	Consump. (m)
Overlocking	Col. I	120/3	250
Sewing	Col. I	120/3	160
	Col. II	120/3	190
Buttonhole	t. i. t.	80/3	90

Shoulder Padding：No. 27

Zipper：Vislon 7#, 1 × 72 cm, t. i. t.

Cord stops：ST56, 4 × t. i t.

Eyelets：middle, 6 × bronze

Nylon biased tape：Art. 25, 3. 2 cm × 460 cm

Embroidery patch："Grace" Col. II

Hanger loop：Shell I, 10 cm (finished lg. 6 cm)

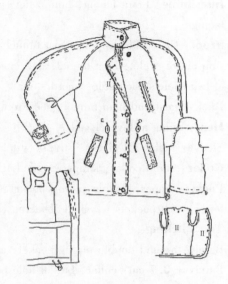

John & Grace Fashion Company

1234 Victoria Street, Hong Kong

Page 2

WORKING SHEET

Season：Winter 2021

Style No.：2564　　　　　　　　　　　　　　**Size**：　8

Description：Ladies' jacket with buttoned-on detachable lining　　**Length**：　80cm

Technical Details：

Jacket：inner body with plain taffeta lining, inner sleeves with quilted taffeta lining, w. buttoned on detachable body lining.

Shell II：detachable body lining, sleeve strap facing, welt facing, button hole panel facing.

40 gr. Wadd.：front facing, back neck facing, collar, button hole panel, detachable body lining.

Quilted taffeta：inner sleeves.

Inner process：back neck facing + front facing turned over, detachable lining bund w. 0. 8 cm nylon bias tape, buttoned on front facing + back neck facing, w. loop at armhole firmly.

Shell string：2 × 0. 8 cm br. × 170 cm;

Button dist.：2. 5/22. 5/22. 5/22. 5/1 × collar strap

Detachable lining：CB/5/5/17/17/17/17/2 × armhole w. loop

Attention：substitute button below on left front facing.

Overlap：

Interlining：Front facing, button hole panel(1×), collar(2×), piping, straps(1×), back neck facing

Front：CF visible zipper, waist tunnel (taffeta), 3 cm br. Understitched w. shell string + cord stop out of eyelet; hem tunnel, 2.5 cm br. by hem turning-over w. shell string secured at CF out of two eyelets each side + cord stop.

Back：waist and hem tunnel, see Front.

Hem：lining at hem closed.

Sleeve：raglan, 2 pieces; straps, stitched on, buttoned.

Collar：stand collar; straps w. shell II facing, buttoned.

Pocket：upper, double piping pocket (2×0.7 cm br.), pocket bags: 1×shell I + 1×taffeta; lower: welt pockets w. shell II facing, pocket bags: 1×shell I, 1×taffeta

Top stitching：

Barely：around double piping pocket, armhole, front edge;

Barely + 0.7 cm：collar, button hole panel + interfacing, upper arm seam, straps

Barely + 2 × 0.7 cm：welts;

2 cm：sleeve hem,

2.5 cm：hem

Barely + 2.5 cm：neckline facing, front facing edges.

Words and Phrases

working sheet	工艺单
specification sheet (SPEC sheet)	工艺单，规格说明书
garment making-up	服装成衣
transaction [trænˈzækʃən]	交易
technical parameter	技术参数
pattern number	样板编号
style number	款号
hard and fast rule	一成不变的法则
working sketch	效果图
specifications	规格，规格说明
yarn count	纱支数
fabric area density	织物面密度
colour card	色卡
colour-matching table	配色表

t. i. t.（tone in tone）	配色
medium [ˈmiːdjəm] size	中档尺码
working drawing	效果图
assembly [əˈsemblɪ]	衣片缝合
stitch type	线迹类型
seam type	缝子类型，缝型，缝式
stitch density	线迹密度
seam neatening	缝边收光
customary [ˈkʌstəmərɪ] practice	习惯做法
interlacing [ˌɪntə(ː)ˈleɪsɪŋ]	线环（圈）交叉
interlooping [ˌɪntə(ː)ˈluːpɪŋ]	线环（圈）互串
intralooping [ˌɪntrəˈluːpɪŋ]	线环（圈）自串
overedge seaming	包缝，包边缝
edge seaming	包缝，锁边缝
unravel [ʌnˈrævəl]	脱散，拆散，散开
single thread chain stitch	单线链缝
basting [ˈbeɪstɪŋ]	粗缝，假缝，疏缝，疏缝的针脚
blind stitch	暗缝
hemming [ˈhemɪŋ]	缲边
hand stitch	手工线迹
hand-stitch machine	仿手工线迹缝纫机
double-pointed needle	双针尖缝针
lockstitch [ˈlɒkstɪtʃ]	锁式线迹
tension	张力
back tacking [ˈtækɪŋ]	回车
double-thread chain stitch（double locked chain stitch）	双线链缝
centre seat seam	裤后裆中缝
cover stitch	覆盖缝
overlock stitch	包缝
to bind [baɪnd]	捆边
trimmed edge	修剪过的布边
two-thread overlock stitch	双线包缝
three-thread overlock stitch	三线包缝
double-needle four-thread overlock stitch	双针四线包缝
safety stitch	安全缝
flat seam stitch	绷缝
seaming	缝合
bridge	跨接

butted join	平接口
interconnect	相互连接
needle thread	缝针线
stitch cycle	线迹形成循环，成缝周期
puckering	缝迹起皱
superimposed seam	叠缝
lapped seam	搭接缝
butted [bʌtɪd] or flat seam	对接缝/平接缝
bound [baʊnd] seam	包边缝
decorative ['dekərətɪv] seam	装饰缝
applied seam	在织物一边装"镶片"的缝子
penetration [penɪ'treɪʃən]	穿透
fashioned knitted garment	成形编织的服装
oversewn	拷边，包缝
binding	滚边
bias-cut ['baɪəs-kʌt]	斜丝缕裁剪，斜裁
pinking [pɪŋkɪŋ] shear	锯齿形布边剪
pinking	锯齿型布边裁剪
non-fraying [nɒn'freɪɪŋ] fabric	不散边织物
double-turned	双道翻折
clean finish	光整处理
self-neatening	缝子自行光边
French seam	法式缝(俗称"来去缝")
measurement tolerance ['tɒlərəns]	测量容差
intermediate and final inspections	中期检验和最终检验
loose-fitting garment	宽松式服装
tight fitting garment	紧身式服装
fictitious [fɪk'tɪʃəs]	虚构的
waist relaxed	放松状态的腰围
waist extended	拉开状态的腰围

Exercises

A. Please use " T " or " F " to indicate whether the following statements are TRUE or FALSE：

(1) A working sheet should clearly specify the style number that the working sheet is related to.

(　　)

(2) A working sheet must clearly specify the material consumption for the specific garment style.

(　　)

(3) If the colour of an accessory is specified as "t. i. t.", it means that the colour of that accessory must be similar in tone to the colour of the shell fabric. 　　　　　(　)

(4) In stitch formation interlacing is where a loop formed by one thread passes through another loop formed by a different thread. 　　　　　(　)

(5) In lockstitches, loops of one (or a group of) thread(s) are passed through the fabrics where they are secured by interlacing with the thread or threads of another group. 　　　　　(　)

(6) Stitches within the Class 300 have poor extensibility and will unravel easily if the thread is broken. 　　　　　(　)

(7) At both ends of the seam formed by stitches of stitch Type 301, back tacking is often used to secure the stitch. 　　　　　(　)

(8) For all garments, seam neatening is always necessary. 　　　　　(　)

(9) On working sheets, all information about the colours involved in a garment must be clearly defined. 　　　　　(　)

(10) All technical details about how to make up a particular garment style must be clearly provided in working sheets. 　　　　　(　)

B. Please select the word(s) or phrase(s) that makes the statement correct:

(1) Stitches widely used in garment production to bind the trimmed edges of a fabric fall within Class_____ 　　　　　(　)
    a) 300 　　　　b) 100 　　　　c) 500 　　　　d) 400

(2) Stitch Type_____is a single thread chain stitch and is extensively used for sewing with temporary stitches. 　　　　　(　)
    a) 101 　　　　b) 301 　　　　c) 202 　　　　d) 401

(3) The stitches in the diagram below are in Class_____. 　　　　　(　)

    a) 300
    b) 400
    c) 500
    d) 600

(4) The fabrics shown below are joined by a_____. 　　　　　(　)

    a) superimposed seam
    b) lapped seam
    c) butted seam
    d) bound seam

(5) _____may need relatively wider seam allowances. 　　　　　(　)
    a) A curved seam line 　　　　b) A straight seam line
    c) Loosely-woven structures 　　　　d) Plain single jersey fabrics

Reading Materials

Spotlight on Sweatshirt

Textiles with subtle shimmer or dazzling shine

Deana Tierney May

It's sweatshirt season — we look at some of the fabrics to make this autumn classic!

Organic

This European-made fabric brand embraces GOTS organic fabric throughout its range. Working with the renowned designer Hamburger Liebe, Albstoffe produce amazing fabrics, together with matching trimmings, all finished to the highest German quality. They have two French Terry sweatshirt and jersey collections — Glow and Wanderlust. Both are 95% organic cotton with 5% elastane and have wonderful butterflies, lilies and safari prints.

Stripes

A Breton stripe top is a classic for any sewers wardrobe. The Malo fabric collection from Modelo is a classic yarn-dyed Breton style stripe on French Terry with every stripe 12 mm wide. This fabric comes in a range of subtle colours so there's bound to be a colour for your project.

Fleece-backed

The Fulton Fleece Backed Sweatshirt is a superb quality, fleece-backed sweatshirt which is very soft to touch. The poly/cotton mix and Oeko-Tex certification mean that it's kinder to the environment during manufacture and it will stay looking great for much longer. It is available in an extensive range of 16 colours.

Heathered

Heather refers to the colour effect that happens when two yarns are interwoven and mixed together. It's typically used to mix multiple shades of grey or grey with another colour to produce a muted shade. The Malmo Heathered French Terry is a looped-back sweatshirt that's soft to the touch and available in a wide range of colours. Many with matching ribbings.

—excerpted from *Modern Sewing starts here*, edition 18-October 2021, page 10.

【参考提示】

1. Sweatshirt 指"长袖运动衫"。国内也有人将其译作"卫衣",应该是来自粤港地区的"卫生衫"称呼。它们常用毛圈布或起绒布缝制,所以也可被称为"毛巾衫"或"薄绒衫"。它们一定是长袖的,通常为圆领,也有带风帽的。在英语中带风帽的常被称作 hoody。

2. GOTS 是 the Global Organic Textile Standard 的缩写,即"全球有机纺织品标准"。

3. Albstoffe 是德国以生产有机织物而闻名的织物生产商。本文出现不少关于品牌和公司的英语专有名词,若无国内普遍接受的中文译名,也不妨暂用英语原词。如 Breton 条纹布(Breton stripe)、Fulton 背面起绒的长袖圆领衫(Fulton Fleece backed sweatshirt)等等。

4. French terry 单面针织毛圈布。一般单面针织毛圈布的工艺反面就是毛圈。本文中 looped-back 就是指用毛圈一面作为服装衣料的反面。如果毛圈一面的毛圈经平剪并刷绒工艺处理，那么就成了"背面起绒"（fleece-backed）。

5. poly/cotton mix 涤棉混纺。

6. Many with matching ribbings 指"许多（Malmo 的麻点夹花单面针织毛圈布）有配色的（可做袖口、领口或下摆的）罗纹布"。

Vogue patterns V1870, Misses´ Jacket and Pants

Semi-fitted, lined, below hip length jacket has notch collar, shoulder pads, princess seams, flaps, welt pockets and long two-piece sleeves with non-working vent and button trim. Semi-fitted, high rise pants have flared legs, stitched front crease, fly zipper, contour waistband with hook and bar closure.

Fabrics：Wool Blends, Linen Blends, Gabardine, Jacquard.

Interfacing：Lightweight Fusible.

Lining：Lining Fabric.

Unsuitable for obvious diagonals.

Note：Fabric requirement allows for nap, one-way design or shading. Extra fabric may be needed to match design or for shrinkage.

Notions：A：One Pair of 1/2″ Shoulder Pads, One 7/8″ (2.2 cm) Buttons Cover, Six 5/8″ Buttons Cover.

B：One Hook and Bar, One 7″(18 cm) Zipper.

Size Combinations：B5(8-10-12-14-16), Y5(16-18-20-22-24-26)

Yardages：

Fabric widths given in inches.										
SIZES	8	10	12	14	16	18	20	22	24	26	
A. Jacket										
44, 45"	$2^{5/8}$	$2^{5/8}$	$2^{5/8}$	$2^{5/8}$	$2^{5/8}$	$3^{1/4}$	$3^{1/4}$	$3^{1/4}$	$3^{1/4}$	$3^{1/4}$	yds.
58, 60"	2	2	2	$2^{1/8}$	$2^{1/8}$	$2^{1/2}$	$2^{1/2}$	$2^{1/2}$	$2^{1/2}$	$2^{1/2}$	yds.
Interfacing										
1 3/8 yds of 20" Lightweight Fusible										
Lining										
44, 45"	$2^{1/8}$	$2^{1/8}$	$2^{1/8}$	$2^{1/4}$	$2^{1/4}$	$2^{1/4}$	$2^{1/2}$	$2^{1/2}$	$2^{1/2}$	$2^{1/2}$	yds.
B Pants										
44, 45"	$2^{1/2}$	$2^{1/2}$	$2^{1/2}$	$2^{1/2}$	$2^{3/4}$	$2^{3/4}$	$2^{3/4}$	$2^{7/8}$	$2^{7/8}$	$2^{7/8}$	yds.
58, 60"	$2^{1/4}$	$2^{1/4}$	$2^{1/4}$	$2^{1/4}$	$2^{3/8}$	$2^{1/2}$	$2^{1/2}$	$2^{1/2}$	$2^{1/2}$	$2^{1/2}$	yds.
Interfacing										

3/4 yds of 20" Lightweight Fusible

—excerpted from https://somethingdelightful. com/vogue-pattern/new-sewing-patterns/_spring/v1870

【参考提示】

1. 该网站是美国 McCall 样板公司官网。这是它发布的 Vogue Patterns® 2022 年春季目录封面的新款时装套装介绍。

2. with non-working vent and button trim 指"带有假袖衩及扣饰"。

3. stitched front crease 指"加缝的裤腿前挺缝线褶痕"。

4. with hook and bar closure 用搭钩和条形搭扣扣合。

5. jacquard 这里指"提花布"。这里 Fabrics 后面列出的四个词语中前两词和原料有关，而后两词和织法有关，不应并列。Wool Blend or Linen Blend, Gabardine or Jacquard 更合理。

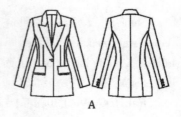

A

6. Unsuitable for obvious diagonals 这里应该是说 This pattern is unsuitable for fabrics with obvious diagonal effects。diagonal 这里指"斜向配置的花纹"，该表达强调了 obvious，因为推荐使用的 Gabardine(华达呢)本身就是一种斜纹布。

7. … nap, one-way design or shading 这里的表达其实就是本书第五章课文中提及的 pile direction 问题，nap 指"绒毛"，design 应该是"花型"的意思。

8. 1/2″ shoulder pads 半英寸厚规格的肩衬。

9. 7/8″ Buttons Cover 这里的 Buttons Cover 应该为 Button Cover，如为复数则为 button covers。button cover 指"纽扣(的金属装饰)卡帽"。不要将 button cover 与 covered button 混淆。前者通常是金属制的，卡帽上可以刻各种装饰花纹，甚至镶嵌"珠宝"，可以将它作为饰品卡在普通纽扣上。后者通常指 wrapped-around button，即"包扣"，一般用和衣服相同面料包裹制成。这里的"7/8 英寸"指的是该纽扣卡帽的直径，如按第三章课文中介绍的纽扣号数的定义，则相当于 35 号。

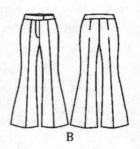

B

10. Size Combinations 服装样板尺码组合，比如 B5 组合是 8 码到 16 码。

11. Interfacing 即 interlining。

Arid Rain Jacket
Weatherproof Technology

Designed to keep the worst of the Elements at bay. Combining Style, Practicality, and Reflectivity, the ETC Arid Jacket range has been conceived with Riders' comfort and safety in mind.

· Waterproof and Fully Reflective on both Ladies' and Men's
· Seam Sealed for added Water proofing
· Deep Tail Splashguard Protection
· Adjustable Cuffs
· Under Arm Vents
· Elasticated Waist Band
· Mesh Inner Liner

- Inside, Side and Chest Pockets
- Fleece Lined Inner Collar
- Machine Washable
- Available in Men's and Ladies'

Arid Nomad Stasher Jacket
Waterproof Technology

One minute the suns shining and then the heavens open. The ETC Arid Nomad Stasher Jacket folds away into its own back pocket to allow it to be stowed easily in a backpack, saddle bag or pocket. Sudden showers needn't spoil your ride again.

- Lightweight Ripstop Fabric
- Highly Waterproof/ Breathable
- Raglan Sleeve
- Fold away hood with toggle and elastic cord
- Folds neatly and easily into its own rear pocket
- Reflective detailing
- Elasticated Cuffs

—excerpted from *ETC* 2021 *Winter Clothing & Accessory Range*, page 3 and 7.

【参考提示】

1. 以上摘自位于英格兰德比市的英国知名自行车及骑手装备经销商 Moore Large 公司 (moorelarge. co. uk) 发布的冬季骑车服装及装备的商品目录。ETC 是 Everything To Cycling 的缩写。

2. keep … at bay 隔绝……。

3. seam sealed 指"经密封处理的衣缝"。通常是在接缝处热压密封胶条做成防水缝, 俗称"压胶"。

4. deep tail 长后摆。

5. splashguard protection 防溅保护。

6. vent 该词与服装下摆、脚口或袖口关联时, 通常指"开衩";与腋下部位关联时, vent 指"排气口"。通常开口处使用网眼布或者还可设计成用拉链开启或关闭。

7. mesh inner liner 网眼布衬里。

8. stasher jacket 可收入自身袋内的夹克衫。

9. the suns shining 应该是想表达"the sun is shining"。其后面的 the heavens … 情况也一样。在网上似乎常出现这种不规范英语的表达。

10. ripstop 抗撕裂的。

11. toggle 相当于 cord stop。

CHAPTER *5* GARMENT MANUFACTURE

1 Introduction to Manufacturing Sequences

Any new garment style comes from the originality of the designer, but only the style that satisfies the market requirements will achieve commercial success. Therefore, good designers will closely monitor social, market and material trends to anticipate and, possibly, influence fashion developments and skillfully design products that will be eagerly received. A good garment supplier should have a reliable design team conversant with current fashion trends and manufacturing methods.

Interpreting and converting a design from either a fashion sketch or a sample garment into a commercially acceptable garment requires its production to be carefully engineered through all the manufacturing stages. Garment manufacturing operations and sequences differ according to the style of garment, but for cut-fashioned garments, as opposed to knit-fashioned garments, the manufacturing follows a similar sequence of processes.

1.1 Manufacturing Sequences for Cut-fashioned Garments

Cut-fashioned garments are those assembled from cut garment pieces.

1.1.1 Pattern Cutting

The first process in the sequence is to interpret the designer's sketch into a set of patterns from which a sample garment may be made. This process is referred to as pattern cutting. Though garment pieces could be produced directly by draping and pinning either actual fabrics or calico onto dress stands, for the most part, in mass production, a set of paper patterns are made. Generally, paper patterns for the medium-sized sample are cut first, and after the samples made with those patterns are approved, patterns for the various sizes needed to fulfill the order are produced from the medium-sized patterns by a process known as grading.

Before cutting the paper patterns, the pattern cutter should carefully study the sketches and style features including any measurements specified by the designer to ascertain the quantity and shape of each pattern piece that is needed. If the paper patterns are developed directly from the buyer's sample, a detailed analysis of the structure of the sample garment is also necessary.

Experienced pattern cutters may use their own formulae to calculate the pattern dimensions. For upper body garment or tops, such formulae are usually based on the bust measurement and some particular measurements, say garment length and sleeve length, or based on the hip measurement and some particular measurements, say the front rise and back rise, for lower garments. Curved or straight structural lines join these points marked on paper to create each pattern.

Alternatively some pattern cutters use basic block patterns to produce the required patterns. Block patterns are patterns produced to a specific shape without any styling features that will fit a different part of the anatomy according to a set of anthropometrical data. Consequently many types of basic block exist including bodice blocks, sleeve blocks and skirt blocks, etc. These block patterns only provide the pattern cutter with a basic pattern shape, and many skills and considerable experience are needed to adapt these patterns into a final set of patterns that correspond to the designer's sketch or sample garment and provide the required degree of fit.

As has been mentioned in Chapter 2, for symmetrical garment pieces such as left and right sleeves or symmetrical left and right front bodies, only one pattern would usually be made initially. Ultimately a complete set of production patterns for each size will normally be produced with a separate pattern for each garment component. For some smaller scale manufacturing operations, when either producing a marker or layout of the pattern pieces to cut the fabric, symmetrical patterns may be used twice, that is, with the face side up to draw one side (say the left) pattern and then with the pattern flipped to draw the other side (say the right) pattern. Similarly, if there is no front opening or no centre back seam, only a half paper pattern to be folded over along the centre front (CF) or centre back (CB) for the front or back body is usually made. "CFF" or "CBF" will be marked near the CF or CB line to indicate the "centre front fold" or "centre back fold".

Notches should be marked on or cut into the patterns to indicate the seam allowances, folding lines, dart sites and balance marks, etc., and grain lines showing the direction in which the pattern should be laid onto the fabric before cutting must be marked. Finally, information about the style number, piece name and size, etc. must be indicated on each pattern as illustrated in Fig 5. 1.

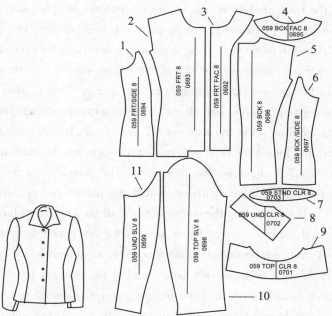

Fig. 5. 1　Sketch of a garment and the patterns required for its shell fabric

1−right front side　2−right front　3−right front facing　4−back neck facing

5−right back　6−right back side　7−collar stand　8−under collar

9−top collar　10−right top sleeve　11−right under sleeve

1.1.2 Sample Making

After the first patterns have been created, samples can be made. Sample making is a very important stage in garment manufacture. As well as providing a sample for a buyer or merchandiser to review, making the sample enables the technical data required for mass production to be derived, such as the fusing duration and fusing temperature, fabric heat shrinkage after ironing, difficult assembly operations to be identified and the total time required for the garment assembly operations to be determined. Such information is very useful for setting up the production parameters. Furthermore, the material and labour costs can be evaluated in sample making.

Samples are made in the sample room, which is usually equipped with the same sewing machines as those in the sewing room. The technicians and workers in the sample room should be highly skilled and experienced. Many manufacturing problems could be detected by the sample room staff during the sample-making stage, thereby making the mass production more efficient and cost effective.

After they have been made, the samples will be sent for approval. If they do not meet with approval, patterns might need correction or modification and samples should be made again until they are accepted by the designer or the buyer.

1.1.3 Grading

In garment production, grading refers to the technique, by which garment patterns are developed from one size to other sizes. If the samples have been approved, the patterns for various sizes to be used in the mass production can be graded from the sample patterns.

X-y coordinates should be defined for each pattern before grading, and grading rules should be devised which define the increments along the x and y directions for every key or cardinal point on the pattern to increase or decrease the size of the master sample pattern to the other adjacent sizes required within the order. The grading rules are usually derived from the grading charts, which, in turn, are derived from the size charts based on anthropometrical data.

Manual grading requires skills and experience, and is very labour intensive because even for one style of garment, the total number of patterns for all sizes may be considerably large.

Computer-aided grading simplifies the procedure. The sample patterns for the medium size are first input into the computer by digitising around the pattern pieces, and, with the help of grading rules stored in the computer, the coordinates for all the key points for each pattern will be automatically calculated, size by size. These key points are joined by the computer, usually through lines created using mathematical spline functions, to automatically create the patterns in the various sizes. If the size range is very large, and if the patterns involved are complicated, garment samples for the smallest size and the biggest size usually need to be produced to ensure that no distortion occurs between the graded patterns. If the master sample pattern has been created directly using the computer, there is, clearly, no need to digitise the pattern pieces into the computer.

1.1.4 Fabric and Accessory Inspection

Fabric inspection should be made for both quantity and quality. After delivery to the clothing factory, fabrics must be inspected often piece by piece carefully. The level of inspection depends on the clothing manufacturer's confidence in the fabric supplier to produce quality fabrics and to identify faults through their own post production inspection. Some clothing manufacturers create a supplier rating based on their experience of previous supplies of fabric. If the supplier is new, the probability that inspection will be done piece by piece is higher. If the fabrics are flat-fold packaged, they need to be wound on to rolls before they can be inspected on the fabric inspection machine. Each piece of fabric should be numbered (tagged) and all data resulting from the inspection be recorded accordingly.

One of the main purposes of fabric inspection is to check whether there are any fabric faults. Tags will be used to mark at the selvedge where faults are found so that these can easily be located during fabric spreading. Fabric widths and piece lengths should be recorded for subsequent reference during lay planning. Where there is width variation within a fabric roll, the narrowest width should be used as the basis for lay planning and marker making. Colour shading on each piece of fabrics needs to be checked. If the piece length is long enough and if the colour tone only has a slight difference along the whole piece length from the beginning to the end, such colour shading could be accommodated during garment production. However, if the shading occurs in short intervals or exists across the width of fabrics, the fabrics might not be useable because colour shading on the garments would be inevitable or fabric utilisation would be very low.

For each lot of fabrics, colour matching may be necessary, especially for piece dyed fabrics made from natural fibres, which are more liable to have piece to piece colour shading. Small swatches are cut from each piece of fabric, with their piece numbers marked. All the swatches are laid closely together on a table and viewed under natural light, and then regrouped according to their colour tones. Those fabric pieces with similar widths and in the same cutting group can be spread together using the same lay-plan.

After fabric inspection, the information on piece length, piece width, numbers of fabric faults and the number of the cutting group, etc. should be recorded. An illustration of this is shown in Fig. 5.2. Fabrics with severe or excessive number of faults should be rejected before cutting, since claims regarding the fabric quality are unlikely to be accepted after cutting.

Inspection of the inherent fabric quality is also very important because the inherent quality of a garment to a great extent depends on that of fabric. According to the specifications set in the fabric purchase contract, fabric specimens are selected randomly and then tested to check for colour fastness and shrinkage, etc. If necessary, appropriate mechanical properties of the fabrics will also be measured. Furthermore, ecological tests should be made to check whether they contain residual heavy metal, formaldehyde or pesticides, etc., and whether they satisfy the criteria set by the importing government.

Likewise, for the accessories to be used, inspection of both quantity and quality are also necessary when they are delivered to the clothing factory.

Shanghai Dongxu Garment Factory

123 Dongxu Road, Shanghai

Fabric Inspection Record

Date: 22 Nov. 2021

Fabric Art. No. :2200

Piece Number	Colour No.	No. of Faults	Piece Width(cm)	Piece Length(m)	Shade Group	Remarks
1	4	2	110	38	1	
2	4	0	110.5	41	2	
3	4	1	110	40.5	1	
4	8	0	111	39.6	3	
5	8	0	110	38.5	3	
6	8	1	110.5	40	4	

Fig. 5.2 Example of a fabric inspection record table

1.1.5 Lay-planning and Marker-making

The "lay" refers to the stack of fabrics that are spread prior to cutting. A "marker" is the layout of pattern pieces that will be used to cut the fabrics in a single lay. Lay planning is the overall plan of how all the lays will be produced and it includes the number of patterns of each size that will be included in each marker, the marker making itself, the number of markers needed to fulfill the order, the number of plies that will be spread in each colour for each of the markers and the length of each lay that is to be spread for each marker.

In lay-planning, the following questions need to be answered:

a) Which sizes should be included in each marker?

b) How should the fabric be spread? For example, how many plies should be spread for each colour and can fabrics of different colours be cut together?

Clearly, the information from both the fabric inspection and the colour/size/quantity assortment from the buyer's order are needed to make the above decisions. Furthermore, the types of the fabrics involved, the working capacity of the cutting machines and the length of the cutting tables, etc. also need to be taken into account.

Consider the simple order shown in Fig. 5.3. which has the quantities for the sizes in each colour equally distributed. A reasonable way to make the markers might be for the smallest and largest size patterns to be arranged in one marker, and the patterns for the second smallest size and the second biggest size to be arranged in another marker and so on. Consequently, if the patterns for sizes 4 and 14 are included in the same marker, and if 40 plies of the marine coloured fabric and 30 plies of the apricot coloured fabric can be spread in the same lay, then all garment pieces for those two sizes can be produced in a single cutting operation.

John & Grace Fashion Company

1234 Victoria Street, Hong Kong

Style No. : 25644

Order No. : H3045

Date : 22 Nov. 2021

COL	SIZE						
	4	6	8	10	12	14	TTL
Marine 52	40	80	120	120	80	40	480
Apricot 46	30	60	90	90	60	30	360
TTL	70	140	210	210	140	70	840

Fig. 5. 3　A garment order with colour/size/quantity equally distributed

Of course, in garment manufacture it is invariably much more complicated than this. The quantities in the various sizes in an order may not be equally distributed, and fabrics, nominally of the same colour, may have to be divided into two or more groups due to total variations in colour or widths. Furthermore, the total number of plies that may be spread in a single lay will be constrained by the thickness of the fabrics and the stroke and length of the cutting blade used. Fabrics with bulky structures would also be restricted in the number of plies that could be spread because the bottom layers might be distorted due to the heavy pressure from the top layers.

The final lay-plans can only be determined after marker making. If a factory has a CAD system that enables markers to be pre-made easily, it may be better, if time permits, to make several plans, according to the widths of the fabrics and the pre-made marker lengths, and then select the best one for the production.

After a decision has been made on which sizes should be included in each marker, the markers can be made. For a clothing factory without a CAD system, marker making has to be done manually. It is a complex process, which usually needs both skill and experience. All paper patterns for the various sizes should first be re-cut in strong and durable cardboard and then the actual marker width, corresponding to the net width of the fabric to be cut, should be marked onto paper. The patterns are arranged one by one, usually starting with the larger ones, to fit them as close together as the various different shapes will allow within the marked area and marked with a pencil around the outlines of the patterns. Using tailor's chalk to draw markers directly onto the fabric is appropriate for single bespoke garment but is not cost-effective for mass production and therefore is seldom used. Furthermore, to draw a marker in such a way will result in imprecise cutting and make it difficult to adjust the marker.

Care must be taken to ensure that the grain lines on each pattern are properly aligned with the length direction of the marker. For some fabrics, certain patterns such as front sides, under sleeves and back sides can be slightly tilted off-grain to improve the fabric utilisation but, when it is allowed, the left and right pattern pieces should be given a similar and opposite tilt to preserve the ultimate matching appearance of the garment components.

Fabric pile direction also has to be taken into consideration. For fabrics with one-way pile such as velvet, all patterns must be positioned so that they all lie in the same direction. For fabrics with a two-way pile, all the patterns within each garment should be positioned in the same direction but the direction of the pattern positions may differ between garments if the design will allow it. For fabrics such as solid chintz or poplin, no directional restriction needs to be applied. The same considerations need to be taken for fabrics with printed, knitted or woven patterns (motifs) (see Fig. 5.4). For example, when fabrics with two-way motif are used, patterns can be positioned on the same principle as the fabric with two-way pile if the garment order allows. It follows that where restrictions are applied on the positioning of patterns the fabric utilisation will be lower.

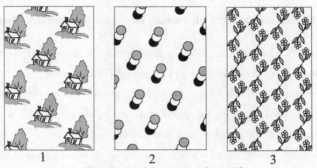

Fig. 5. 4   Direction of motifs
1−one-way motif   2−two-way motif   3−"non-directional" motif

Furthermore, for fabrics with printed, knitted or woven patterns, especially with checks or stripes, pattern matching may be necessary to ensure that the line of strips or checks across the front panels (in particular) align. Poor matching of the checks or stripes at the patch pocket, side seam, centre back seam and front opening, etc. would create a poor visual effect.

In manual marker making it is very difficult to have a clear overview of the whole marker being made, and it is also not easy to re-position the drawn patterns, and consequently, an experienced worker is needed who knows where patterns need to be positioned to maximise fabric utilisation; otherwise fabric utilisation could be lower, which in turn will increase the production cost. In some factories without a CAD system, markers are made first by positioning a full set of patterns, previously miniaturised usually in 1/5 scale, within a proportionally reduced area. In this way, the whole marker can be observed and adjusted. After the miniaturised marker has achieved a satisfactory marker length, the formal full size marker to be used for production can be drawn with the original patterns according to the arrangement within the miniaturised marker.

Nowadays, most clothing factories use CAD systems, and the marker making process becomes much simpler. The patterns for the particular sizes stored in the computer are called and listed or displayed at the top of the monitor screen in the form of icons. On the centre of the screen, the marker width is marked and the icons, starting with those representing the larger pattern pieces, are dragged and positioned on to the marker in a suitable order until all the pieces have been positioned. The computer system will automatically align the grain lines unless instructions allowing the rotation of patterns are given. If there are restrictions on fabric pile direction or pattern direction, the operator can input the instructions on such restrictions and the computer will position the garment patterns according to these instructions. The system will show the length of fabric used and the fabric utilisation for each marker, and the latter is usually calculated from the actual area of the patterns themselves and the product of the marker width and marker length. The patterns are positioned and re-positioned until an acceptable marker length is reached.

The marker is now ready for plotting. It is advisable to plot the marker only when it is needed, especially in conditions of high humidity, because humidity could cause the length of the plotted marker to change. In computerized cutting, the marker data is directly sent to the cutting tool, controlling the latter to move and cut the spread fabrics.

Fig. 5. 5 illustrates a cutting marker, in which only one garment is involved. Generally, if more garments are included in a marker, the fabric utilisation would normally be higher but much depends on the sizes included in the marker, the shape of the pattern pieces and the relative proportion of large and small pattern pieces. Of course, the length of a cutting marker will be constrained to that of the cutting table.

Fig. 5. 5　Example of a cutting marker for patterns shown in Fig. 5. 1
(**No pile restriction imposed**)

1. 1. 6　Fabric-spreading

Fabric spreading is made according to the instructions from the lay planning. Fabrics with the relevant piece number are moved to the cutting table, and spread according to the instructed marker length and mode of spreading. Some fabrics can only be spread in one direction, that is, the fabric will be cut at the end of each traverse and no fabric is spread in the return traverse. Some fabrics can be spread in two directions, in other words, there is no need to cut the fabric at the end of traverse and fabrics can be spread to and fro, which may lead to a higher productivity in spreading.

During spreading, the tension on the fabric needs to be even and as minimal as possible since it could cause garment pieces to contract after relaxation or even distort after cutting. Furthermore, one edge of the fabric layers must always align to ensure that all garment pieces are subsequently cut correctly. If there is a fabric fault, the fabric can be cut from the point where the fault occurs, in which case it is necessary, when recommencing spreading, to overlap the fabric to ensure that the garment pieces across the cutting line are complete. Alternatively, the fabric layer with the fault may be removed for later use and spreading recommences at the beginning of the lay.

For knitwear, especially knitted underwear, fabrics usually are spread and cut in a tubular form instead of in open width and their cutting markers tend to be much simpler.

1.1.7 Cutting

After fabrics are spread, the plotted or drawn marker is laid on the top of the fabric plies (or "lay" as it is called). Following the outlines of the patterns, the garment pieces are cut. Experience and skill are still necessary to ensure that the cut edges are even and vertical to the cutting plane. There is no hard and fast rule about how to start the cutting but many operators prefer to cut the small pieces first.

Inspection of the cut pieces is necessary and if a certain garment piece is found to be distorted or incorrectly cut, a replacement should be cut. Care should be taken to ensure that the new piece should be of the same colour tone to that of the one replaced.

1.1.8 Sorting, Bundling and Numbering

After cutting, the stacks of garment pieces for the same garment in the same size are sorted and put together either in bundles of the same components or in sets that include all the pieces needed for a single garment. The former way is used in the bundle production system and those garment pieces would be delivered manually for sewing in bundles. The latter way is used in the unit production system and the sets of garment pieces would be delivered to individual workstations, according to the operation that needs to be performed, by an overhead, computerized and mechanized chain handling system.

To avoid possible colour shading, it is often necessary to number the garment pieces. The garment pieces from the same layer should be numbered with the same number to remind the operators that the garment pieces cut from different fabric layers should not mistakenly be joined together.

1.1.9 Fusing

For some garment pieces such as collars, pocket flaps and front facings, interlinings should be attached to them to reinforce them before sewing. The commonly used interlining in the clothing industries is the fusible interlining which can be stuck, using heat and pressure, on to the relative garment pieces through a fusing process.

Temperature, pressure and time are the most important factors in fusing. Sufficient heat is needed to change the dry thermoplastic fusible adhesive into a partially molten state, and sufficient pressure needs to be applied to ensure intimate contact between the fusible interlining and the garment pieces. Too low temperature or too low pressure would lead to low adhesion with possible de-lamination during laundering, dry cleaning or wear. Too high temperature or too high pressure would result in excessive penetration of the adhesive into the garment pieces, giving "strike back" or "strike through" effects with resultant poor handle, adhesion, wear and cleaning properties and performance. The time should be sufficient to permit the temperature and pressure to achieve the required melting and penetration effect for the adhesive of the interlining. Of course, these three factors are interrelated, and the appropriate temperature, pressure and time should be determined when the sample garment is being made.

1.1.10　Sewing, Under Pressing and Trimming

After the garment pieces have been cut, sorted and numbered if necessary and after the necessary interlinings have been applied, the garment components can be delivered to the sewing shop for garment assembly.

Production lines should be organized according to the sequence of assembly operations needed for each style of garment. It is beyond the scope of this book to give an account of assembly operations for different styles of garments as there are many steps involved in the garment making-up process. Raw edges of the garment pieces may need to be neatened especially for unlined garments; garment pieces must be assembled with appropriate stitch and seam types according to the instructions from the working sheets and labels should be attached during the assembly stage. There are some operations in which, for example, the top collar and under collar with attached interlining are joined with right side against right side, and then turned inside-out (referred to as "bagging out"). Under pressing is necessary before the collar can be sandwiched onto the bodice.

Similar manipulations will be adopted when making pocket flaps and waistbands. Hems may need to be turned up and secured using hemming attachments and bar tacks would be applied to secure the edges of openings. Templates may be used to mark the positions of buttons and darts.

After the garment components are joined together to form a garment, the assembled garments need to be trimmed to remove the unnecessary thread ends, and then, inspected. Those garments with sewing faults will be returned to the work place that originated the fault, whereupon piece replacement may also be necessary, which would require an urgent request to be sent to the cutting shop.

1.1.11　Finishing and Final Pressing

After inspection and rectification (where necessary), assembled garments will be sent to the workshop for final pressing.

Some garments such as jeans and denim jackets might need washing. There are many types of garment washes, including sand washing, stone washing and enzyme washing, but such processes are often performed in special finishing factories. Some garments may need to be printed or painted. Hand painted kimonos are an example of the latter and such work may be subcontracted if the skilled workers and equipment are not available in the clothing factory.

The final pressing is also called top pressing. The purpose of final pressing is to remove all creases on the garments to present them attractively ready for packaging or presentation on hangers. Some types of accessories that are susceptible to damage under the high temperature and high pressure used during the final pressing may need to be attached after garment final pressing.

1.1.12　Garment Inspection

Finally, the garments are ready for the final inspection. For clothing factories, such an inspection is generally made on the apparent quality of the garments. The apparent quality of a garment may include the workmanship, fabric faults, colour shading and measurement. The inspection is usually made to use random sampling, or, piece by piece, if requested.

Where the garment contract is agreed based on the quality of the original sample or the counter sample, this sample is used as the criterion to determine whether or not the garment being inspected should be rejected. If the garment contract is not agreed by sample and if there are clear specifications in the contract, the inspection is undertaken according to the contractual quality terms.

Garments will be measured to check whether, for example, the bust girth, waist girth, garment length and sleeve length conform to the data in the size chart.

Where specified, hangtags will be attached to garments according to buyers' instructions. The rejected garments need to be closely re-examined to see whether the rejection is justified, and, if the garments warrant rejection, remedies are usually sought and applied unless it is really impossible to rectify the faults.

1.1.13　Packaging

Garments may need to be packaged or put on hangers. Although the garment sales contract may specify packing terms, many buyers may only provide detailed packaging instructions before production starts. The instructions would also be contained in the working sheets and they may comprise the following details:

a) the way how the garments should be folded;

b) the packed dimensions of the folded garments;

c) the size/colour/quantity assortment in each package;

d) the packing materials and their method of application;

e) the package dimensions;

f) marking on the packages.

After they are packed, the garments are now ready for dispatching. Packing lists should be made out to record the actual packing details.

For those garments to be transported on hangers in containers, each garment is encased in a polybag and hung on rods in the container. Strings are used to secure the hanger hooks to prevent the hangers from sliding off the rods during the transportation. The containers are usually loaded at the factory premises, and then, transported to the container yard for shipment.

1.2　For Knit-fashioned Garments

Some of the processes for knit-fashioned garments are similar to those used in manufacturing the cut-fashioned garments. However, since the garment pieces are directly formed from yarns on knitting machines into the required shapes, the processes of pattern designing, grading, lay planning, marker making and cutting, etc. are not needed.

After they are delivered to the factory, yarns need to be inspected to check whether their physical and visual properties conform to their specifications. Then the approved yarns are sent to the knitting workshop. Where the yarns for knit-fashioned garments are delivered to the factories in hanks, they will need winding onto packages. The winding is generally done twice, that is, winding from hanks to cones and then rewinding from cone to cone. The purpose of winding in this way is to ensure an even yarn tension in the yarn packages. Modern winding machines have either sophisticated defect detection devices or simple slub catchers at each winding head to detect and remove faults and imperfections in yarns. Oiling or waxing may be necessary to make the yarns softer, to reduce their friction and to prevent electrostatic charges from interfering with knitting.

Before the garment pieces are knitted, all technical details must be engineered carefully according to the designer's instructions and sketches, or according to the samples from the buyer, together with the measurement specifications for each size of the finished garments.

Prior to volume production, a sample is knitted to determine and verify the knitting sequences needed and to produce a sample for approval. This involves several processes. The appropriate machine gauge (the number of needles per inch along the needle bed) should first be selected according to the thickness or linear density of the yarns to be knitted and the required tightness of the knitted structures. Then according to the maximum width of a garment piece and the wale density specified, the number of working needles, including those required for seam allowances, can be calculated. According to the length of the garment piece, the stitch types knitted by the needles for the particular structures and the course density, etc., the number of carriage traverses can be deduced. Finally, the amount and frequency of the widening and narrowing needed are programmed according to the shape of the garment piece. In this way, the working details for all garment pieces can be determined and they will be listed in the working sheet with necessary instructions. Diagrams or working sketches can be used to make the knitting specifications clearer.

Most garment pieces are knitted on normal V-bed knitting machines but some may be knitted on fully-fashioned machines, which are programmed to produce the required designs and structures using the combinations of knit, tuck and miss stitches. Yarn colours and loop transfers, etc. specified in the knitting instructions are also programmed directly using the CAD system, which is used to create the designs as well.

Following the instructions given by the working sheet, all garment pieces can be knitted piece by piece. Before the pieces are joined together, they may need to be relaxed to relieve inherent stresses and strains imparted in the knitting process−this is usually done on steaming machines. Inspections of garment pieces are also necessary.

Garment pieces are assembled either on linking machines (stitch against stitch in the case of precision linking) or on cup seaming, flat seaming, overlocking, chain stitching and lock stitching machines. The latest technology in V-bed weft knitting enables complete garments to be knitted. These require only minimal sewing operations and thereby greatly reduce the amount and cost of the assembly operations needed.

Both the assembled and whole-knitted garments may undergo a series finishing processes according to the specifications, such as milling or fulling, moth-proof finishing, shrink-proof finishing or pill-resistant finishing, and then the garments are washed, dewatered, and dried. A final steam setting is often performed which is usually done by the steam press.

In the final inspection, the finished garments are inspected, usually through random sampling, to verify the garment measurements and check got faults, loop density and garment weight, etc. The parameters that should be inspected generally depend on the contractual terms or original samples.

Packaging of the knit-fashioned garments should follow the packing specification agreed during the business negotiation. In most cases, each garment is packed in a polybag, and then boxed according to specific colour/size/quantity assortments and subsequently packed in cartons.

2 Equipment in Clothing Factories

In modern clothing factories machines are used to perform as efficiently as possible the tasks needed at each operation. In addition to the design area, a clothing factory may have three main working areas, divisions or rooms, i. e. the cutting room, sewing room and finishing room. The design area houses the sample making room which invariably has duplicates of those machines used in the sewing room to enable the operations needed for each stage of production to be engineered to be as efficient as possible. Many of the machines are purposely built for specific operations and also to de-skill the sewing operation and reduce product variability introduced by variation in operators' skill and performance.

In many factories the pattern cutter also works in the sample room, using the tools for making patterns, such as pattern master, French curves, metre rules, tracing wheels, shears, pattern notches, tape measures and pencils. Dress stands are also necessary in the sample room for trying on the finished samples or for developing patterns.

Some factories have CAD centres. In addition to computers, printers and plotters, a CAD centre should be equipped with a digitizer, with which paper patterns can be input into the computer system. Scanners and digital cameras may also be used to facilitate garment or pattern designing. Clearly, the tasks can only be accomplished on the computer with the necessary software installed for pattern design, grading, and marker making, etc.

A clothing factory should also have a fabric inspection room, which is usually near the warehouse. In that room, there should be one or two fabric-inspection machines, which offer enough illumination so that the inspectors can readily detect the fabric faults as fabrics are unwound and rewound over inspection tables. Some machines can also measure the fabric length and width, and have facilities for tagging the positions of faults.

2.1 Equipment in Cutting Rooms

2.1.1 Cutting Table

In the cutting room fabrics are spread on the cutting table, and then, cut, sorted, bundled and numbered if necessary. The cutting table should be sturdy and smooth, with sufficient length and width to provide cost-effective fabric spreading and accommodate the range of fabrics to be spread. Some cutting tables are equipped with vacuum devices to hold the fabric down or air-blowing devices to float the fabric to enable spreading and cutting to be effectively performed. Those tables used for cutting may also be able to have their height conveniently adjusted to facilitate cutting.

2.1.2　Spreading Machine

There are many types of fabric spreading machines or cloth spreaders as they are often referred to, the simplest of which may only be a fabric carriage mounted on the cutting table. Typically they comprise a carriage supporting the fabric wound on a roll which unwinds over a tension roller, also mounted on the carriage, when the carriage moves along the table. Such spreading machines are usually operated manually, but there are some spreading machines which are driven by electric motors and the rotation of the fabric rollers on those machines can also be driven by motor to ensure that only a minimum tension is evenly exerted on the fabric.

On a fully automatically programmable spreading machine, all the main spreading functions are stored in the memory of the microprocessor. The data on lay length and the type of spreading, e.g. one-way face up, one way face to face or two-way zigzag, can be keyed in with a keyboard. Edge detectors can send signals, controlling the carriage to move laterally, to ensure good edge matching for all plies of fabrics. They may also have circular blades that move across the carriage to cut the fabric at the end of each traverse.

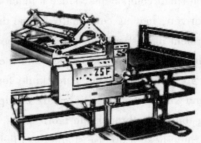

Fig. 5.6　Spreading machine

2.1.3　Cutting Tools

Unlike in the sample room, in which fabrics are cut with tailor's shears because usually only a single ply of fabric is to be cut, most cutting tools used in the cutting room are electrically, hydraulically or pneumatically powered equipment. Of all the cutting equipment, the straight knife cutter (or straight knife, see Fig. 5.7) is the most commonly used in clothing factories. On the straight knife cutter, the cutting knife is driven in a reciprocating movement by an electric motor mounted on the top. Since the blade of the knife is usually around 20 millimetres wide, it is often difficult to cut precisely curved parts such as the neckline and armhole on the garment components. The knife is heavy and is supported on rollers beneath a platform which passes under the plies of fabric to be cut. The difficulties in cutting are compounded by the weight and vibration of the motor, which make it difficult to manipulate. However, it is very versatile and can cut a wide range of fabric types. There is a handle on the straight knife. When cutting the fabric, the operator holds the handle and gently pushes the cutter along the outlines of patterns in the marker laid on top of the spread fabrics. The ply height of the fabrics being cut could reach more than twenty centimetres, but this usually depends on the specific type of cutter.

The rotary or round knife cutter (or round knife, see also Fig. 5.7) is another cutter which is driven by an electric motor and operated by hand. However, since the diameter of the circular knife may not be very large, the ply height that may be cut is usually around 20 to 70 millimetres and it is difficult to be used to cut sharp corners.

Fig. 5.7   Straight knife (1) and round knife (2)   **Fig. 5.8   Band knife**

The band knife (see Fig. 5.8) is specifically used to cut the curved parts and sharp corners on garment pieces, especially on small components, since a much finer metal band can be used as the blade. The band is power driven, moving continually in one direction, and this makes cutting more stable. Generally, the garment pieces with curved parts to be cut precisely are first separated with a straight knife, leaving the curved parts untrimmed, and then the plies of "roughly" cut or "blocked-out" garment pieces are moved onto the band knife for precision cutting. Unlike the previous two knives, on the band knife, the fabric plies are manipulated past the cutting head by the operator rather than moving the blade through the fabric plies.

For those garment pieces which need a precise shape in sufficient quantity, die cutting can be used. There are mechanical die cutters and hydraulic die cutters. Although the die cutter can offer precisely cut garment pieces, their relatively higher cost and the cost of producing the dies may preclude them from being used widely. Die cutters could be used for garment pieces such as collars, pocket flaps and motifs to ensure uniform shapes.

Modern technology has made computerized cutting possible. Similar to a plotter, under the control of computers the cutting head, which is usually a reciprocating knife but in some limited applications could also use a laser or water pressure, etc., moves through the spread fabric, supported on a bed of easily deflected nylon bristles, to cut out the garment components piece by piece. Perforated paper is usually used to spread first below the bottom ply of each lay so that the lay can be kept well supported and the vacuum on the cutting table can work effectively as well. A sheet of airtight polyethylene is used to cover the top of the lay to facilitate the vacuum compressing the lay tightly during cutting. Despite the much higher cutting speeds compared to the manually operated cutting tools, the high capital cost of such equipment necessitates large volumes of production to justify the initial expenditure.

2.1.4 Fusing Equipment

As mentioned in the previous part of this chapter, the fusible interlining is widely used in clothing production. There are several types of continuous fusing machines to apply the temperature and pressure needed, but usually they have one or two heating sections, a pair of pressing rollers and a conveying system. On the conveyor belt of the fusing machine, the garment pieces to be reinforced with interlining are placed piece by piece, face down with the corresponding cut piece of the interlining laid on top of them. The belt then transports them continuously through the heating sections and between the pressing rollers (see Fig. 5. 9). The adhesives (resin) on the interlining melt and fuse the interlining onto the garment pieces.

　　In addition, discontinuous fusing equipment could be found in some clothing factories. These are referred to as flat-bed fusing presses. Among these is the steam press, which is normally used for intermediate or final pressing of the garment (see Fig. 5. 10). The garment piece and the cut piece of the interlining are placed together and pressed between the platens (or "bucks") of the steam press. Heat is provided by steam and the temperature is dependent upon the steam pressure available. Pressure is usually applied mechanically or pneumatically and a timer on the press can control the pressing cycle.

　　A steam iron may be one of the simplest fusing tools, but the fusing temperature, pressure and time applied would vary from piece to piece, and consequently the fusing quality is not easy to control and cannot be assured.

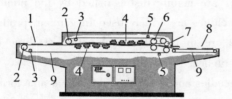

Fig. 5. 9　Illustration of a continuous fusing machine

1—loading station　　2—tension roller
3—control roller　　4—heating element
5—cleaning bar　　6—pressing roller
7—strip-off device　　8—cooling station
9—conveyor belt

Fig. 5. 10　Illustration of steam press

1—control panel
2—top platen
3—assembly to be fused
4—bottom platen

2. 1. 5　Other Equipment in Cutting Rooms

　　There may be some other equipment in cutting rooms to meet special technical requirements. One example is the numbering gun or labeling gun, which can be used to mark the same number on those garment pieces cut from the same ply of fabrics. The number can indicate to the sewing operators those garment pieces (i. e. those with the same number) that need to be assembled into the same garment. Another example is the strip-cutting machine, which features multiple circular knives to produce straight or bias strips. Those strips can be used for binding or piping, etc.

2. 2　Equipment in Sewing Rooms

　　The most important equipment in sewing rooms, without doubt, is the sewing machine. However, there are many types of industrial sewing machines, which can be classified by their stitch type, number of needles involved, machine speed, application, or degree of automation. A fully automatic sewing machine is only used for some very simple operations, and most sewing operations have to be performed by human operators.

2.2.1　Overview of Primary Sewing Machines

The lockstitch machines are widely used in clothing factories. The straight stitch structure of the lockstitch machine is suitable for joining almost all fabrics except those which have a great deal of elasticity. If classified according to the number of needles with which they are equipped, there are usually two types of lockstitch machines, i. e. the single-needle lockstitch and the double-needle lockstitch machines.

The chain stitch machine is more suitable for use in joining those fabrics which have elasticity because the loop formation in its stitches provides the seams with both extensibility and elongation strength. There are several types of chain stitch machines that produce single-thread, double-thread or multi-thread chain stitches, using single, two or even more needles.

The overlock machine is used primarily for trimming and neatening the raw edges of the fabrics and for joining some knitted or other stretch fabrics. The overedge loop formation of the overlock stitches can completely enclose the fabric edges and at the same time provide the seams with good extensibility. Overlock machines are mainly distinguished by the number of threads involved. Three or four-thread overlock machines are widely used in garment production. Seams from five or six-thread overlock machines are in fact the combination of three or four-thread overlock stitches and double-thread chain stitches.

Flat seaming machines feature multi-needles and multi-threads. The seams formed by the flat seaming machines have sound extensibility, good strength and neat appearance and they are widely used for attaching knitted rib cuffs or necks and joining the shoulder seams of knitwear, etc.

2.2.2　Primary Systems in a Sewing Machine

Generally speaking, there are four primary systems in a basic sewing machine. These are the needling system, the thread hitching system, the thread take-up system and the fabric feeding system.

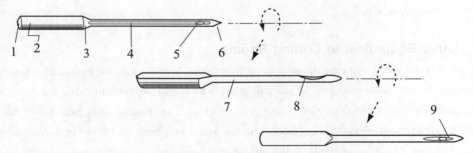

Fig. 5.11　Sewing needle

1—butt　2—shank　3—shoulder　4—long groove　5—needle eye

6—needle point　7—needle blade　8—scarf　9—short groove

The chief component in the needling system is the needle, which is usually straight but may be curved on some sewing machines. The needle is specially designed so that it can perform well at the very high operational speeds of between 5000 – 8000 stitches per minute. Referring to a needle used on an ordinary lockstitch machine (see Fig. 5. 11) as a typical example, it has a strong shank to be held by the needle clamp on the machine. On one side of the needle blade there is a long groove which protects the needle thread as the needle penetrates the fabrics and ensures that the thread does not bulge out on that side of the needle as it withdraws from the fabrics. On the other side there is a short groove which also protects the thread as the needle descends through the fabrics. There is a scarf or cutaway section at the needle eye to facilitate the point of the rotating hook (or other sewing element in the case of other stitch types) passing between the needle and the thread to catch the thread and form the stitch. There are generally two types of needle points, that is, round points and cutting points. The former are commonly used for sewing textile materials and the latter are used for sewing leather or similar strong materials. Within each type there are many sub-types, which are used according to the type of fabric being sewn, to minimise damage to the fabric.

The thread hitching system (or the stitch forming system) ensures that the thread loop can be formed and interlaced properly by the thread catcher. On the lockstitch machine, the rotating hook together with the hook point works as the thread catcher. On other machines, such as the chain stitch or overlock machine, loopers and spreaders may be used (See Fig. 5. 12. Please note that presser foot and feed dog, etc. are not shown).

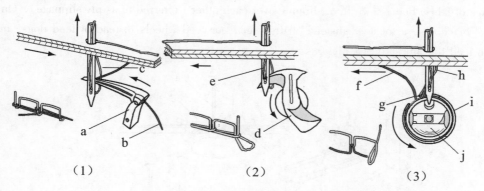

（1）　　　　　　　　　　　　（2）　　　　　　　　　　　　（3）

Fig. 5. 12　Examples of thread hitching system

（1）-for double thread chain stitching：a）looper　b）looper thread　c）needle thread

（2）-for single thread chain stitching：　d）rotating looper　e）needle thread

（3）-for lock stitching：　f）bobbin thread　g）point of hook　h）needle thread

i）rotating hook　j）bobbin case

In the thread take-up systems, a take-up lever or a take-up cam is widely used. The function of the thread take-up system is to pull the thread out of the packages as the needle descends so that enough thread can be supplied to form the thread loop, and to pull the excessive thread back as the needle ascends so that balanced stitches can be formed (See Fig. 5. 13).

The function of the fabric feeding system is to move the fabrics forward between each stitch formation cycle. There are several types of feeding systems. The most common system is the drop feed mechanism, which comprises a set of serrated metal teeth (referred to as feed dogs) that protrude and descend through a slot in the throat plate that supports the fabrics. A spring loaded presser foot presses the fabrics being sewn against the feed dogs and the rotary movement of the feed dogs under the fabrics pushes the fabrics forward. The horizontal displacement of the feed dogs determines the distance moved by the fabric between each stitch and hence the stitch length.

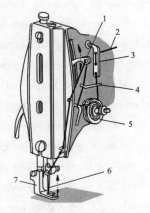

Fig. 5. 13   Illustration of take-up lever, tension disc, etc.

1-take-up lever   2-needle thread
3-thread retainer   4-thread guard
5-tension discs   6-needle   7-presser foot

Differential feeding, using two sets of feed dogs, can be used to produce a stretching or gathering effect. Reciprocal feeding or compound feeding, whereby the needle (and sometimes the presser foot) may also move horizontally whilst the needle is penetrating the fabric, may be used on some machines to prevent the plies of fabric being sewn from slipping over each other (referred to as ply slippage). On some types of machines, a top feed instead of drop feed, or feed wheels instead of feed dogs, may be used to feed the fabrics (See Fig. 5. 14).

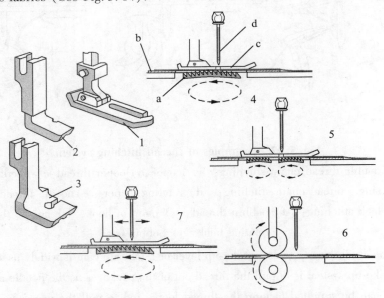

Fig. 5. 14   Examples of presser foot and feeding mechanism

1-hinged presser foot with two toes   2-single piping foot   3-double piping foot
4-drop feed   5-differential feed   6-wheel feed   7-compound feed
a) feed dog   b) throat plate   c) presser foot   d) needle

In addition to these systems, some auxiliary devices can be fitted to a sewing machine to perform some special functions. One good example is the piping foot which can facilitate the making of cord piping.

2.2.3　Other Equipment

There are also other kinds of sewing machines in a sewing room. Typically for many garments, buttonhole machines are needed and they can be classified as either straight buttonhole machines or eyelet buttonhole machines. Associated with these are button-sewing machines, which are designed to attach two or four-hole plain circular buttons of various diameters using principally chain stitches. Some are designed to have special attachments so that they can also attach shank buttons and some snap buttons.

The felling machine (sometimes called the blind stitch hemming machine) sews chain stitches by a curved needle such that the stitches are almost invisible on the right side of the garment.

Bar tacks are often used to secure the ends of garment openings, such as pocket openings and vents, for which the bartacker is the right machine to serve this purpose. The bartacker can also be used to attach labels. There are some other machines engineered for special operations, such as jig sewing (sometimes called template sewing, see Fig. 5.15) machines and quilting machines. The former enables garment components such as pockets, cuffs and collars to be made accurately and efficiently, and the latter are designed specially for the quilted garments with wadding. Some clothing factories are equipped with embroidering machines, which are necessary to embroider garments or garments with embroidery patches.

Fig. 5.15　Jig sewing

Most modern sewing machines incorporate needle position motors which are programmable so that they sew for a specific number of stitches and stop automatically. On stopping, the needle may be either in the pre-set raised or lowered position and the presser foot automatically lifted. Sensors may be included to detect the end of the fabric and stop the machine accordingly.

Many sewing workstations are designed around these semi-automatic machines to de-skill the sewing operation and they work in conjunction with stacking devices to automatically stack the sewn components. Such machines are widely used for products that show few style variations and are produced in large quantities such as shirts and jeans. Examples of such workstations include machines for automatically setting pockets, run-stitching collars and seaming pocket flaps.

2.2.4　Production Lines

Unlike in a tailor's shop, where a garment is typically assembled by only one person, in mass production many people would be involved in the sewing room to make up a garment. There could be several types of sewing operations required even for a single component of a garment and it can be easily understood that different stitch types would have to be produced on different sewing machines. Furthermore, even though collars and sleeves may be made up using a lockstitch machine, to achieve the volume of production needed, several machinists in the sewing room would be required: some to prepare collars and some to prepare sleeves. Similarly, the prepared collars and sleeves would be attached to the garment bodies by another group of machinists, even though the stitch type involved is still the lockstitch. The number of machinists depend on the time required for the individual operations and the skill of the operator. This gives rise to the concept of a standard minute value for an operation. The need to balance production to ensure a smooth flow of work through the sewing room is of paramount importance. Balancing the production is achieved by taking into account the standard minute value for each operation and the performance of the operator. The repetition of the same work by the same operator makes that person much more skilful and faster in carrying out the operation and hence improves his/her performance. Since, in the sewing room, most operators are paid by piece work, i. e. their pay is directly proportional to the number of operations that they perform, the operators are keen to increase their skill level and operational speed.

For the majority of factories, the production systems used are the progressive bundle system and the unit production system or a mix of both, with the progressive bundle system being the most commonly used.

In a factory adopting the bundle production system, each machinist completes a single or series of operations on a bundle of garment components and then sends the bundle to the next station for further processing. In a factory adopting a unit production system, a special garment handling system conveys hangers containing all the garment components for a single garment between workstations at which each operator will perform an operation and, the hanger will automatically be dispatched to the next workstation immediately after the operation is done (See

Fig. 5. 16   Unit production system

Fig. 5. 16). The unit production system gives a faster throughput in that there is no waiting for the rest of the components to be sewn in a bundle before it is dispatched. However, it is less tolerant on imbalances in production between workstations than the bundle system and, as such, apart from the capital investment on the conveyor system, it requires careful real time production monitoring to ensure its efficiency. As previously mentioned, hybrid systems have been developed whereby some sub-assembly operations are done using a bundle system and the final assembly is done in units. Garment components are often transported by less sophisticated manually driven chain hanging rail systems.

How to organize different types of machines to form a production line is a question which is worthy of careful consideration. Generally, it can be done as follows: firstly, the stitch types required for the garments need to be established; secondly, the corresponding machines need to be chosen; thirdly, the sequences of different operations need to be determined; fourthly, the production tasks allocated to each workstation need to be calculated to minimise production imbalances, and finally, the layout and organisation of the machines need to be planned to minimise the time spent on handling or transporting unfinished garments from one workstation to another.

2.3　Equipment Used in the Finishing Room

The most common garment finishing operation performed in the clothing factory is the final pressing. Special finishing operations such as garment washing, e. g. sand washing, stone washing or enzyme washing, are usually subcontracted to factories with the specialised equipment and expertise needed.

The simplest pressing equipment used in the finishing room is the iron which is usually a steam iron with constant steam supply. Steam irons with an electric heating element embedded in them can provide a better pressing effect than ordinary steam irons. Associated with irons are ironing tables that need to be sturdy and covered with the heat-safe padding, and (ideally) have vacuum-extraction to remove the heat effectively after ironing so that the garments can be handled safely. The disadvantage of garment pressing with irons is that the pressing quality wholly depends on the skill of the operators.

The steam press, which may also be used for fusing, can give a much better top pressing quality. The temperature, pressure and pressing time can be pre-set. Uniform and constant quality can be expected. Various types of steam presses can be found in the finishing room of a clothing factory to more efficiently and consistently press some special parts or serve special purposes. These include shirt presses, cuff and collar presses, trousers or jacket presses, etc., some of which are arranged as carrousels and have devices to automatically remove the pressed component.

In addition, form finishers for jackets (Fig. 5. 17) and shirts, etc., cabinet finishers (Fig. 5. 18), tunnel finishers (Fig. 5. 19), thread removal machines and stain removing machines, etc. could be found in the finishing sector of some clothing factories.

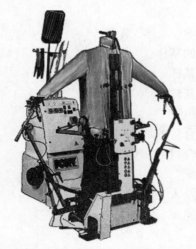

Fig. 5. 17　Form finisher for jackets　　　　**Fig. 5. 18　Cabinet finisher**

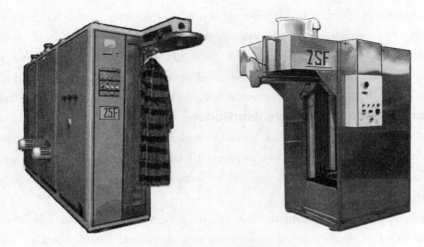

Fig. 5. 19   Tunnel finisher

Words and Phrases

manufacturing sequences	生产流程
originality [əˌrɪdʒɪˈnælɪtɪ]	创意
be conversant with [kənˈvɜːsənt]	对……熟悉的
fashion sketch	时装效果图
engineer	工艺设计
pattern cutting	样板制作，出服装样板，打板
calico [ˈkælɪkəʊ]	白棉布，本白布
dress stand	胸架，人形模型
pattern cutter	样板设计师，打板人员
structural line	结构线
basic block pattern	基样，原型
anatomy [əˈnætəmɪ]	解剖学，人体，骨骼
bodice block	大身的基样
sleeve block	袖的基样
skirt block	裙的基样
dart site	省的位置
fusing duration	上黏合衬的加热时间
fusing temperature	上黏合衬的加热温度
heat shrinkage [ˈʃrɪŋkɪdʒ]	热收缩
material and labour costs	材料及人工成本
sample room	打样间
sample-making	打样
sample pattern	样衣的样板

x-y coordinate [kəuˈɔːdinit]	x-y 坐标
computer-aided grading	计算机辅助推档
spline [splain] function	样条函数
fabric inspection	织物检验
piece by piece	逐匹地
flat-fold packaged [ˈpækidʒid]	折叠式包装的
fabric inspection machine	验布机
selvedge [ˈselvidʒ]	布边
fabric spreading	铺料
fabric width	织物门幅
piece length	匹长
lay planning	铺料设计
width variation [ˌveəriˈeiʃən]	门幅变化
marker making	排板，排料
colour tone	色调
in short interval [ˈintəvəl]	短片段的
colour matching	配色
piece dyed fabric	匹染的织物
natural fibre	天然纤维
swatch [swɒtʃ]	样布
claim	索赔；主张
inherent [inˈhiərənt] quality	内在质量
purchase [ˈpɜːtʃəs] contract	购货合同
randomly [ˈrændəmli]	随机地
mechanical property	机械性能
ecological [ˌekəˈlɒdʒikəl] test	生态测试
residual [riˈzidjuəl]	滞留的，残余的
heavy metal	重金属
formaldehyde [fɔːˈmældihaid]	甲醛
pesticide [ˈpestisaid]	杀虫剂
lay	铺好的布层
marker	排料图，排板图
colour/size/quantity assortment	颜色/尺码/数量搭配
working capacity [kəˈpæsiti]	生产能力
cutting table	裁剪桌
marine [məˈriːn]	海军蓝
apricot [ˈeiprikɒt]	杏黄
stroke [strəuk]	工作动程
cutting blade	裁刀

bulky structure	膨松结构
distorted	变形，走形
net width	净门幅
tailor's chalk	划粉
bespoke [bɪ'spəʊk] garment	定制的服装
tilt [tɪlt]	(使)倾斜
off-grain	偏离丝缕线
pile direction	绒毛方向
one-way pile	单向绒毛
velvet ['velvɪt]	天鹅绒，丝绒
two-way pile	双向绒毛
solid	素色的，单色的
printed, knitted or woven pattern	印花的、针织的或机织的图案
motif [məʊ'tiːf]	图案
pattern matching	对花
miniaturised ['mɪnɪətʃəraɪzd]	缩小的
icon ['aɪkɒn]	图标
drag	拖拽
plotting	(绘图机)绘制
humidity [hjuː'mɪdɪtɪ]	湿度
traverse ['trævə(ː)s]	(铺料设备)横移
tubular ['tjuːbjʊlə] form	圆筒状的
open width	开幅的
bundle production system	"成捆"作业方法
unit production system	"单元"作业方法
workstation	工位
overhead, computerized and mechanized	架空的、计算机控制的、机械
chain handling system	式的链条式搬运系统
fusing	上黏合衬
resin ['rezɪn]	树脂
de-lamination [dɪˌlæmɪ'neɪʃən]	分层，脱层
strike back	黏合剂渗穿织物
strike through	黏合剂渗穿织物
handle	手感
sandwich	夹(着)缝(上)
neaten	(毛边)收光
under pressing	中间烫，小烫
hemming attachment	缲边装置
bar tack	套结，加固缝

template	模板
trim	修整，修剪
thread end	线头
sewing fault	缝纫疵点
finishing	整理
final pressing	成衣熨烫
jeans	牛仔裤
denim jackets	牛仔布夹克
garment wash	成衣洗水
sand washing	砂洗
stone washing	石洗
enzyme [ˈenzaɪm] washing	酵素洗
hand painted kimono [kɪˈməʊnəʊ]	手绘和服
subcontract [sʌbˈkɒntrækt]	合同转包，外包
top pressing	成衣熨烫
workmanship [ˈwɜːkmənʃɪp]	做工
random sampling	随机取样
original sample	原样
counter sample	对等样品（通常指卖方按买方原样做的样品）
contractual quality terms	合同的品质条款
remedy [ˈremɪdɪ]	补救
rectify [ˈrektɪfaɪ]	调整
hanger [ˈhæŋə]	衣架
packing list	装箱单
to make out...	缮制……
polybag	塑料袋
container	集装箱
hanger hook	衣架挂钩
factory premise [ˈpremɪs]	厂区内
container yard	集装箱堆场
hank [hæŋk]	纱绞
winding [ˈwaɪndɪŋ]	络纱
yarn package	纱的卷装
cone	（圆锥型）纱筒，纱的筒子
defect	疵病
slub catcher	清纱器
imperfection	缺陷
oiling [ˈɒɪlɪŋ]	上油
waxing [ˈwæksɪŋ]	上蜡

electrostatic [ɪˈlektrəuˈstætɪk] charge	静电荷
volume [ˈvɒljuːm; (US)-jəm] production	批量生产
machine gauge [ɡeɪdʒ]	机号
linear [ˈlɪnɪə] density	线密度
wale density	纵行密度(横密)
working needle	参与工作的织针
course density	横列密度(纵密)
carriage	机头游架
V-bed knitting machine	横机
fully fashioned machine	全成形机
loop transfer	移圈
stress	应力
strain	应变
relax	松弛
steaming machine	蒸汽定型机
linking machine	套口机
stitch against stitch	线圈对线圈
cup seaming machine	(针织毛衣用的)包缝机
flat seaming machine	绷缝机
overlocking machine	包缝机
chainstitching machine	链缝机
lock stitching machine	锁式缝机(俗称"平缝机""平车")
milling	缩绒
fulling	缩绒
moth-proof finishing	防蛀整理
shrink-proof finishing	防缩整理
pill-resistant finishing	防起球整理
dewater [diːˈwɔːtə]	脱水
steam press	蒸汽压烫机
loop density	线圈密度
to de-skill	降低技能要求
pattern master	打板专用尺
French curves	曲线板
metre rule	米尺
tracing wheel	滚线轮
shears	剪刀
pattern notches	(样板)刀眼夹钳
tape measure	软尺，皮带尺
printer	打印机

plotter	绘图机
digitizer ['dɪdʒɪtaɪzə]	数码转换板
scanner	扫描仪
digital ['dɪdʒɪtl] camera	数码相机
fabric-inspection machine	验布机
vacuum device	吸风装置
air-blowing device	吹风装置
fabric spreading machine	铺料机
cloth spreaders	铺料机
keyboard	键盘
edge detector	探边器
electrically powered equipment	电动设备
hydraulically [haɪ'drɔːlɪkəlɪ]	
powered equipment	液压传动设备
pneumatically [nju(ː)'mætɪkəlɪ]	
powered equipment	气动设备
straight knife cutter (straight knife)	直刀型(电)裁刀
rotary cutter	圆刀型(电)裁刀
round knife cutter (round knife)	圆刀型(电)裁刀
band knife	钢带式(电)裁刀
die cutting	用模具冲裁
die cutter	(模具)冲裁刀
nylon bristle ['brɪsl]	尼龙刺须
perforated paper	打孔纸
continuous fusing machine	连续式黏合机
heating section	加热区
pressing roller	轧辊
conveying system	传送系统
conveyor belt	传送带
fusing machine	黏合机
discontinuous [ˌdɪskən'tɪnjʊəs]	
fusing equipment	非连续式黏合设备
flat-bed fusing press	平板式黏合机
loading station	上料台
control roller	控制辊
heating element	加热元件
cleaning bar	清洁杆
tension roller	张紧辊
strip-off device	剥离装置

cooling station	冷却台
control panel	控制面板
top platen	上压板
bottom platen	下压板
timer	定时器
numbering ['nʌmbərɪŋ] gun	打号枪
labeling gun	打标签贴的枪
strip-cutting machine	开滚条机
sewing machine	缝纫机
lockstitch machine	锁式缝缝纫机(平缝机)
single-needle lockstitch machine	单针平缝机
double-needle lockstitch machine	双针平缝机
chain stitch machine	链缝机
overlock machine	包缝机
needling system	刺料系统
thread hitching system	钩线系统
thread take-up system	挑线系统
fabric feeding system	送料系统
butt	针柄头
shank	针柄
shoulder	针肩
long groove	长针槽
needle eye	针眼
needle point	针尖
needle blade	针身(针杆)
scarf	针缺口
short groove	短针槽
bulge [bʌldʒ]out	凸起
the point of rotating hook	旋梭梭尖
round point	球形针尖
cutting point	棱边针尖
stitch forming system	成缝系统
thread catcher	勾线器
rotating hook	旋梭
hook point	梭钩尖
looper	弯针
spreader	线叉
rotating looper	旋转钩针
bobbin ['bɒbɪn] thread	梭子线,底线

bobbin case	梭壳
take-up lever	挑线杆
take-up cam	挑线凸轮
thread retainer [rɪ'teɪnə]	过线板
thread guard [ɡɑːd]	导线钩
tension disc [dɪsk]	张力盘
presser ['presə] foot	压脚
stitch formation cycle	成缝循环
drop feed mechanism	下送料机构
feed dog	送布牙
throat [θrəʊt] plate	针板
differential [ˌdɪfə'renʃəl] feeding	差动式送料
reciprocal [rɪ'sɪprəkəl] feeding	往复送料
compound ['kɒmpaʊnd] feeding	复合送料
ply slippage ['slɪpɪdʒ]	缝料滑移
top feed	上送料
feed wheel	送料滚轮
piping foot	嵌线压脚
cord piping	嵌线
hinged [hɪndʒd] presser foot with two toes	双趾铰式压脚
single piping foot	单嵌线压脚
double piping foot	双嵌线压脚
buttonhole ['bʌtnhəʊl] machine	锁眼机
straight buttonhole machine	平头锁眼机
button-sewing machine	钉扣机
felling machine	缲边机
blind stitch hemming machine	暗缝缲边机
bartacker ['bɑːˌtækə]	套结缝纫机，打结机
jig sewing	模板缝纫
template sewing	模板缝纫
quilting machine	绗缝机
embroidering [ɪm'brɔɪdərɪŋ] machine	绣花机
embroidery patch	绣花贴片
needle position motor	缝针定位马达
sensor	传感器
stacking ['stækɪŋ] device	堆布装置
machinists	缝纫工
standard minute value	标准测时值
performance	绩效；性能

to be paid by piece work	计件付酬
production line	生产线，流水线
progressive bundle system	逐捆作业方法
handling system	搬运系统
throughput [ˈθruːpʊt]	产量
capital [ˈkæpɪtl] investment [ɪnˈvestmənt]	资本投入
real time production monitoring [ˈmɒnɪtərɪŋ]	实时生产监控
hybrid system	混合作业法
sub-assembly operation	次级的成衣操作，成衣中间的操作
ironing [ˈaɪənɪŋ] table	烫台
heat-safe padding	耐热衬垫
shirt press	衬衫压烫机
cuff press	袖口压烫机
collar press	压领机
trousers press	裤子压烫机
jacket press	上衣压烫机
carrousel [ˌkærʊˈzel]	旋转(木马)式
form finisher	人体模型整烫机
cabinet [ˈkæbɪnɪt] finisher	柜式整烫机
tunnel [ˈtʌnl] finisher	隧道式整烫机
thread removal machine	清线头机
stain removing machine	去污迹机

Exercises

A. Please use "T" or "F" to indicate whether the following statements are TRUE or FALSE：

（1）Grading rules are increments along the x and y directions for every key point on the patterns.

（　　）

（2）Fabric inspections are usually made on the cutting table before cutting is started.　（　　）

（3）The garment sizes to be included in each marker and the mode of fabric spreading will be decided in lay-planning.　（　　）

（4）"Numbering" after cutting is not necessary for fabrics which show no colour shading.

（　　）

（5）"Strike back" caused in fusing means that the back of the fabric being fused melts through excessive temperature.　（　　）

（6）"Winding" before knitting is necessary to provide suitable yarn packages with an even yarn tension　（　　）

（7）The function of the scarf near the needle eye is to facilitate the point of the rotating hook passing between the needle and the thread to catch the thread and form the stitch.　(　　)

(8) The machine gauge of a knitting machine refers to the number of needles in a unit length (usually per inch) along the needle bed.　(　　)

(9) During fabric spreading, excess or uneven tension is undesirable.　(　　)

(10) Where there is width variation within a fabric roll, the largest width should be used as the basis for lay planning and marker making.　(　　)

B. Please select the word(s) or phrase(s) that makes the statement correct:

(1) The maximum number of plies that can be spread in a single lay will be constrained by_____.　(　　)

    a) the thickness of the fabrics

    b) the stroke and length of the cutting blade

    c) whether the fabric is bulky

    d) the colour of the fabric

(2) The purpose of sample making is_____.　(　　)

    a) to make a sample garment for approval

    b) to derive the technical data for mass production

    c) to evaluate the material cost

    d) to evaluate the labour cost

(3) Which of the followings are essential in pattern grading?　(　　)

    a) Counter samples

    b) Patterns for the approved samples

    c) Grading rules

    d) Computer

(4) Which of the following parameters/features will typically be checked during fabric inspections?　(　　)

    a) Fabric width

    b) Fabric piece lengths

    c) Weaving or knitting faults appearing on fabrics

    d) Colour shading

(5) The process of "under pressing" happens_____.　(　　)

    a) during sewing

    b) during cutting

    c) after garments are made up

    d) when interlinings are being fused

Reading Materials

Loop Control®

The innovative needle geometry for perfect loops

Loop Control® with lockstitch

With conventional geometry

Sewing of very dense or very hard materials leads to crushing of the thread between the sewing fabric and the scarf edge during the needle down stroke.

With particularly thick multi-filament threads, there is also a risk of the threads getting hooked on the edge of the scarf:

· Over-twisting of the thread above the sewing fabric and partial untwisting of the thread below the needle plate.

· Negative effects of this twist shifting on loop formation.

With Loop Control® geometry

The improved Loop Control® geometry of the blade and scarf edge provides greater protection of the thread and reduces the load on the thread when it passes over the scarf edge.

Result:

Better thread protection and reliable loop formation; Reduction of skipped stitches and torn threads (due to poor loop formation) even in critical applications.

Loop Control® for chain stitch

With conventional geometry

On 2 to 4 needle chain stitching machines, the needle closest to the looper forms the smallest needle thread loop-due to the short needle rise. Machines which are set to a tight stitch setting or are using textured sewing threads tend to produce skipped stitches.

With Loop Control® geometry

Improved geometry of the blade and scarf and larger eye in comparison to the needle size: Processing of textured sewing threads is improved.

Result:

In combination with the correct machine settings, Loop Control® chain stitching needles form clean and stable loops, even with tight stitch formation and when using textured yarns.

—excerpted from www. groz-beckert. com/mm/media/en/web/pdf/Loop_Control. pdf

【参考提示】

1. 这是摘自 Groz-Beckert 公司官网的关于缝纫针几何形状优化以控制线环形成的 Loop Control®介绍。该公司是老牌的世界知名针织机、缝纫机用针供应商。

2. over-twisting 过分加捻。Untwisting 退捻。

3. skipped stitch 跳线。

4. textured sewing thread 变形丝缝线。

M-SERIES
Multi-ply Automated Fabric Cutting

The Pathfinder M-Series® range of automated fabric cutting machines are known for precision, very high productivity and reliability and lowest power consumption. All 16 models (4 cutting width options × 4 cutting heights) are easily configured to suit most cutting applications. And thanks to clever design, Pathfinder cutters require very little maintenance to continually deliver exceptional cut quality over a long working life.

Advanced CNC oscillating knife cutting technology

There are 16 Models available in the M-Series range of automated fabric cutting machines. Discover how an automated fabric cutter will maximize your productivity!

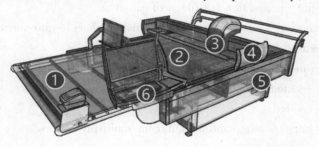

1–Offload Area　2–Cutting Area　3–Cutting Head
4–Safety System　5–Control System　6–UX(User Experience)

Loaded with Standard Features

The M-series cutting system features groundbreaking technologies to ensure your productivity is boosted and profits are maximized.

- Windows® operating system
- Path Cut® cutter operating software
- Touch Screen user interface
- Utility drawer
- Digital vacuum control
- Vacuum re-sealing blind
- Dynamic vacuum monitoring & regulation
- Vacuum exhaust filter
- Vacuum Economiser® power saving function
- QuickDraw® automatic diamond sharpening system
- Automatic bristle cleaning
- Notch identification and transformation
- Drill hole identification and transformation
- Automatic cutting window calculation— 'Whole piece cutting' where possible
- Zoom & scroll graphic finger gesture's
- Piece & cut order identification

· End Cut—operator programmable

· Cutting small pieces first—user definable

· Automatic knife & sharpening stone wear compensation

· Pre-set cutting parameters

· Integration, automatic cut job queuing, cutting list status & completed marker list

· Marker manipulation features, notch selection, rotation, flip, drill omission, scaling etc.

· Conveyor advance pause—automatic user definable

· Automatic conveyor load assist—distance & time user definable

· Automatic pause at piece selected

· Selection of pieces—do not cut

· Automatic lay width checking

· System history file—production cut file verification

· Maintenance manager—automatic reporting of maintenance requirements and schedule

· Head parking—user definable

· Units display in metric, imperial/ yards and inches

· Flaw avoidance—automatic piece movement avoiding material flaws

· Multi-language sported

—excerpted from https://seo. pathfindercut. com/multi-ply-cutters/

【参考提示】

1. 以上摘自澳大利亚的知名布料裁剪机器制造商 Pathfinder Australia Pty Ltd 公司官网。可以注意到，这里仍然采用"关键词堆砌"的方法，简单地直接列出该设备的主要特征。

2. CNC 即 Computerized Numerical Control。

3. oscillating knife 即"振动刀"，它是利用刀具每分钟上万次的上下高频振动实施裁剪。

4. utility drawer 多用途抽屉。

5. Vacuum re-sealing blind 真空再密封帘板。

6. Vacuum exhaust filter 真空排气管过滤器。

7. diamond sharpening system 金刚石磨刀刃系统。

8. bristle 这里指（裁床上的）尼龙刺须，俗称"刚毛"。

9. Zoom & Scroll graphic finger gesture's 指图片缩放及滚动。

CHAPTER *6* GARMENT QUALITY ISSUES

1 The Importance of Garment Quality

It goes without saying that quality is an essential pre-requisite for the acceptance of a product into the market.

Generally speaking, the use of consistently high quality materials, excellent management, skillful operators and well maintained equipment, etc. are factors required for the production of good quality garments. However, it should be noted that quality issues are also closely related to the production costs. To produce garments acceptable to the market at a reasonable cost is an aspiration for many garment suppliers, which means that the quality commercially pursued is usually the acceptable quality.

2 The Criteria for Garment Quality

The level of quality which could be considered as "acceptable" depends primarily on the product specification. Major factors are the need to comply with contractual specifications and governmental regulations. The first criterion, which makes the garments acceptable to the buyer, is set between the buyer and the supplier during business negotiations and the second criterion, which could make the garments acceptable to the market, is set in the statutes defined by the governmental legislature (usually, in international trade by that of the importing country). There should be some tolerances for quality specifications; otherwise, they are either very often unworkable or unaffordable. If the quality of garments always tends towards to the upper limit within the quality tolerance, the garment supplier would be considered to be a good supplier; if it always tends to the lower limit, the reputation of the supplier would be in doubt even if the quality is still acceptable.

In the business negotiation, the garment supplier should think carefully about the contractual terms regarding the garment quality, and must be sure that the quality terms are workable at the cost calculated when making a quotation. Reasonable tolerances should be agreed, and the methods and samples for testing and inspection should be agreed on the basis of mutual understanding about the constraints imposed on both parties. The garment supplier should appreciate that the garment quality is defined in the buyer's working sheets, size charts and/or garment samples. The specifications should be checked carefully and clarification should be sought in cases of uncertainty. For example, the measurement locations and the measurement tolerances in the size chart should be checked to ensure that they are clearly defined.

The garment supplier should also ensure that the company database is kept up to date with regard to governmental legislations regarding garment trade. All garments exported must meet the importing country's local mandatory or regulatory standards and legislation. These regulations are generally related to the inherent quality of garments and defined with the purpose of providing better protection for the environment and the wearer's health and life.

Furthermore, the supplier should know that in Article. 35 (2. b) of *the United Nations Convention on Contracts for the International Sale of Goods*, *1980*, it is stipulated that "except where the parties have agreed otherwise, the goods do not conform with the contract unless they are fit for any particular purpose expressly or impliedly made known to the seller at the time of the conclusion of the contract, except where the circumstances show that the buyer did not rely, or that it was unreasonable for him to rely, on the seller's skill and judgment". In other words, if the buyer wants to buy raincoats, the coats supplied should be rainproof even though there may be no detailed contractual terms regarding the fabric quality.

3 Common Garment Quality Problems

Garment quality may be sub-divided into two aspects: inherent quality and apparent quality. It is necessary to be familiar with typical quality problems and understand what causes these problems.

3.1 Inherent Quality

Inherent garment quality is referred to as the garment quality which cannot be directly observed with the eyes. This is, to a great extent, related to the quality of fabrics, and therefore, if the quality of the fabrics is kept under strict control, it is reasonable to expect good inherent garment quality. For a garment supplier, this would require a certain amount of fabric inspection to be performed. Tests on the fabrics would also need to be undertaken, following the standards expressly or tacitly agreed upon. An experienced supplier would even send his staff to the fabric manufacturing factory to closely monitor the fabric quality.

The following common problems are representatives of those concerning inherent garment quality:

a) Colour fastness. Colour fastness is a sensitive issue to the consumer. No one likes the colour of his or her coat to fade on prolonged exposure to sunlight, and no one is willing to tolerate a beautiful seat cover stained with the colour of the occupant's trousers. There are various types of colour fastness, such as colour fastness to light and colour fastness to wet rubbing, which are defined in terms of grades. The tests conducted usually depend on the contractual quality terms. The higher the grade number, the better is the colour fastness. In many cases, fabrics of natural fibres in dark colours such as black or red tend to have poorer colour fastness. The types of dyes, textile materials involved, the dyeing operations and different setting of dyeing parameters affect the final colour fastness.

b) Shrinkage. Shrinkage is another sensitive issue. Shrinkage arising from steam pressing can be detected during the garment sample making and should be prevented through careful setting of pressing or other parameters. For example, if it is found in making the sample that fusing an interlining to a garment front facing could cause the facing piece to shrink by 2% in the length, the relative paper pattern could be increased by 2% in the length to compensate for this if the fusing conditions remain unchanged. Potential washing shrinkage is unlikely to be known by a customer when purchasing a garment. Although shrinkage might be inevitable even though some pre-shrinkage processes may have been applied, a large shrinkage in the garment or different shrinkages between the lining and the shell will no doubt lead to complaints by customers. Controlling the fabric shrinkage, using appropriate technology and providing appropriate instructions on the care labels, etc. are all actions which should be taken by the garment supplier to minimise potential shrinkage problems.

c) Safety. The safety of a garment concerns the issue whether the garment being worn would affect the wearer's health or life. It should be noted that many countries have issued strict regulations concerning garment safety, and therefore it is important for the garment supplier to know about such regulations.

Some problems regarding the safety may only be critical to certain groups of people. For example, poor colour fastness could affect an infant's health as it is more likely to lick its clothes, and poor flame resistance could easily put infirm or invalided people into danger if their clothes catch sparks from a fire.

Some safety issues will be critical to all wearers. Such issues are mainly caused by the residuals of some harmful contaminants within the garments. Residual pesticides, formaldehydes and heavy metals on the fabric, if they are beyond acceptable limits, would no doubt affect the wearer's health. Scientists have also found that fabrics dyed with 22 aromatic amines that may be reductively cleaved from azo dyes may cause skin cancer. That is why azo dyes are prohibited now by garment importers. Furthermore, it was also found that some dyestuffs such as CI acid Red 26, CI basic Red 9 and CI Disperse Blue 1 are carcinogenic, and that some dyestuffs such as CI disperse Yellow 1 and CI disperse Orange 1 will cause contact dermatitis.

Scientific research has already revealed the following facts:

Chromium (VI) and its compounds are classified as carcinogens but they are quite likely to be found in some heavy-duty textiles. Cadmium and its compounds are also carcinogenic but they can be found in plastisol prints, PVC and PU fabrics. PCP (Pentachlorophenol) and TeCP (2, 3, 5, 6-Tetrachlorophenol) are used as pesticides or fungicides to prevent mould (caused by fungi) but both PCP and TeCP are very toxic and are regarded as cancer-inducing substances. Chlorinated organic carriers such as PCB (polychlorinated biphenyl) and PCT (polychlorinated terphenyl) are mainly used as pesticides and also as softeners, carriers and flame retardants. It has been found that they can easily accumulate in organisms and the environment and can affect the human liver, hormone, immune and nervous systems. Tributyltin (TBT) can be used for anti-microbial finishing but high concentrations of TBT are toxic because it can be taken up via the skin and may affect the nervous system.

It is therefore advisable that the above substances are either forbidden or kept under strict control in garment manufacture.

In addition to factors due to the fabrics, poor garment designs or poor management in garment making-up could also lead to hazards for wearers and one good example is the broken needle left in a finished garment.

There are also some other factors that can lead to unsafe garments. Small accessories or accessory parts on children's garments could be a choking hazard, and therefore, it is advisable to design children's wear without small parts and to ensure that the accessories will only be pulled off with a reasonably large force (e.g., 90 Newtons as mentioned previously). Furthermore, non-durable buttons of wood, cork, leather, mother of pearl (shell) and glass, etc. should not be used on children's garments.

d) **Mechanical properties.** One of the important mechanical properties is the strength of both the fabric and the seams. Little attention needs to be paid to mechanical strength for common items of apparel if suitable fabrics or threads are chosen and appropriate technical procedures are used, but specific attention must be paid for special garments. For some jeans, the bursting strength at the knee would need to be tested and for some types of workwear, the tearing strength would need testing. Therefore, whether or not the mechanical strength should be controlled and what kind of strength tests should be performed will depend on the contractual terms and the intended use for the garment.

3.2 Apparent Quality

The apparent quality of a garment is referred to as the garment quality which can be assessed from the appearance of the garment. It is probably the main factor that a customer considers when he or she chooses a garment. In a clothing factory, the main purpose of the garment inspection after the final finishing process is to examine the garment for apparent quality. There are many types of apparent quality problems, which may be due to the fabric apparent quality or due to poor workmanship or deficient manufacturing skills.

3.2.1 Apparent Quality Problems due to Fabric Faults

Fabric faults in a garment would no doubt affect the garment appearance. Fabric faults may result from imperfect operations in knitting, weaving or dyeing and finishing, or may result from the poor quality of fibres or yarns.

Dyeing faults such as colour spots and colour shading, and knitting or weaving faults such as oil spots, holes and streaks are common fabric problems affecting garment apparent quality. However, it should be noted that it might be possible to prevent these and other fabric faults from appearing on the garments through careful monitoring during the various stages of garment production. Any fabric faults should have been detected during fabric inspection, and, if the fabric faults are likely to manifest themselves in the final garment, tags may be attached to the selvages of the fabric being inspected to remind the operator to take action to remove the faulty length of fabric during the spreading operation. Unless the colour shading on fabrics occurs in short intervals along the piece length or appears from left to right across the fabric width, it may not necessarily be a problem. By ensuring that the fabric pieces of the same colour tone are included within the same lay-plan and using the garment pieces from the same ply to assemble a garment, colour shading may be prevented from appearing in a garment to a great extent.

Whether or not a small fabric fault appearing on a garment would cause the garment to be rejected invariably depends on its location on a garment. Three significant levels for different locations on the garment have been defined. The first level is the most significant level, and this group includes the upper part of the front and back bodices, top collar, upper part of the front facing, top sleeves, front and back of trousers (except for the parts above the upper hip girth and the parts along the inseam). The third level is the least significant to the garment appearance. It includes the back of the lapel, the lower part of the front facing, under the collar, under the sleeves and that part of the trousers near the crotch. The remaining parts of garments fall into the second significant level, and this group would include the part below the waist on an upper garment and the part above the upper hip line on a lower garment. Whether or not a garment with a small fabric fault located in any of the locations at the second significant level mentioned above is acceptable would more or less depend on how that garment is to be dressed. For example, a tiny fault appearing around the shirt hem which is to be tucked in the trousers, or appearing around the waist of trousers which is to be covered by the outer garment top might be accepted by many customers.

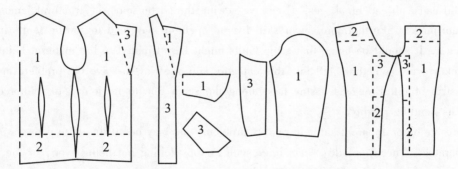

Fig. 6.1   Significant levels for apparent garment quality

3.2.2 Apparent Quality Problems due to Poor Workmanship or Manufacturing Skills

The common apparent quality problems that fall into this category are:

a) Apparent garment problems due to patterns. Generally speaking, if a garment sample is approved, the patterns for a medium size of garment are defined. If a whole lot of finished garments do not conform to the measurements specified, it is necessary to check the patterns. Incorrect grading rules adopted may be a prime cause. It is advisable to make garment samples for both the largest and the smallest sizes after grading. Subsequently, measurements for all locations could be checked according to the size chart and, furthermore, the garment appearance should be examined to ensure that grading on the curved parts of a garment, say armholes and necklines, are perfect. If only a few of garments do not conform to the specified measurements, this may be due to faults in the sewing operations, possibly by the operator who did not strictly follow the seam allowances indicated by notches.

b) **Apparent garment problems due to marker making, spreading and cutting.** If the check or stripe patterns on garment pieces are not correctly matched or if the drape of the left side and right side of a garment are not satisfactorily balanced, there could be something wrong with the marker. The former is due to inconsiderate marker making and the latter may be due to poor alignment of the grain lines of garment pieces. Uneven and excessive tension applied during fabric spreading may also cause relaxation shrinkage which could affect garment measurements. Poor cutting skills could cause cut edges not to be perpendicular to the cutting table and this, no doubt, could also affect the size of the cut parts and hence garment measurements.

c) **Apparent garment problems due to sewing.** There are several kinds of common sewing faults such as skipped stitches, thread breakages, irregular stitches, damage to fabrics and seam puckering, which would affect the garment appearance.

Sewing faults may arise from the improper setting of the machine parts. For example, improperly setting the thread tension or incorrectly selecting or setting the needle could cause thread breakages, and insufficient pressure on the presser foot could cause irregular stitches. Sewing faults may also arise from defective machine parts. Burrs on the needle or on the throat plate could cause thread breakages. Excessive sewing thread tensions or structural jamming may cause seam pucker. Furthermore, poor skill by the operator may lead to sewing faults including seam pucker. It should be noted that some faults might be attributed to the material being sewn. Seam pucker could sometimes fall into this category and one way to solve the problem may be to attach a strip of interlining as a stay tape along the seam line to be joined to reinforce it and increase its bending rigidity.

There are other apparent garment quality problems that may be attributed to poor sewing skills of the operators. These include: seam lines such as topstitching not being even, or the left and right of a garment not being balanced, or sleeves veering towards the front or back, or the top collar being set too tight, etc. Some of these problems could be remedied by skillful ironing operations but others may not. Consequently, new workers should be well trained before being allowed onto the production lines.

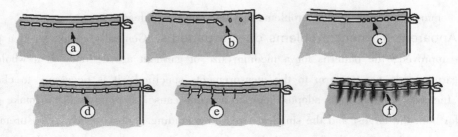

Fig. 6.2 Some typical sewing faults

a) skipped stitches b) thread breakage c) irregular stitches

d) unbalanced interlacing e) damage to fabric f) seam pucker

d) **Apparent garment problems due to ironing.** Garments are usually ironed between sewing operations (called under pressing) or on completion of sewing operations (called top pressing). Poor ironing skill or incorrect setting of the ironing temperature would leave ironing marks on the garment. Glazed or yellowish spots appearing on a garment after ironing tend to be the most common ironing faults. Insufficient pressing or overpressing may also occur because the ironing time, temperature, pressure and the steam applied wholly depend on the operator's experience in hand ironing. Therefore if it is possible, it is better not to use an iron for the final garment pressing because the final quality of the garments will depend on the skill of the ironing operators.

It must be borne in mind that the cause of the apparent problems sometimes might often be complex. For example, a wrinkle in a certain part of a garment may be caused by differential shrinkages between shell and lining, or by poor workmanship, etc. Consequently it is essential in the early stages of garment manufacture to find the exact cause and then devise appropriate remedies. In the context of garment quality, remember that prevention is better than a cure.

Words and Phrases

contractual specifications	合同规格说明
governmental regulations	政府的规章
governmental legislature [ˈledʒɪsˌleɪtʃə]	政府的立法机构
governmental legislation [ˌledʒɪsˈleɪʃən]	政府的立法
United Nations Convention on Contracts for the International Sale of Goods (CISG)	联合国国际货物销售合同公约
apparent quality	外观质量
expressly or tacitly	明示或暗示地
colour fastness	色牢度
colour fastness to light	光照色牢度
colour fastness to wet rubbing	湿摩擦色牢度
dyes	染料
dyeing parameter	染色参数
flame resistance	阻燃性
residual [rɪˈzɪdjʊəl]	残留物
harmful contaminant [kənˈtæmɪnənt]	有害的污染物
aromatic [ˌærəʊˈmætɪk] amine [ˈæmiːn]	芳香胺
azo [ˈæzəʊ] dye	偶氮染料
skin cancer	皮肤癌
CI acid Red 26	颜色指数 26 的酸性红
CI basic Red 9	颜色指数 9 的碱性红

CI Disperse Blue 1	颜色指数 1 的分散蓝
Carcinogenic [ˌkɑːsɪnəˈdʒenɪk]	致癌物(质)的
CI disperse Yellow 1	颜色指数 1 的分散黄
CI disperse Orange 1	颜色指数 1 的分散橙
contact dermatitis [ˌdɜːməˈtaɪtɪs]	接触性皮炎
chromium [ˈkrəʊmjəm](VI)	六价铬
compound [ˈkɒmpaʊnd]	化合物
carcinogen [kæːˈsɪnədʒən]	致癌物质
heavy-duty textile	厚实的纺织物
cadmium [ˈkædmɪəm]	镉
plastisol [ˈplæstɪsɒl] prints	增塑溶胶印花
PVC and PU fabric	聚氯乙烯和聚氨酯织物
PCP (Pentachlorophenol [ˌpentəˌklɔːrəʊˈfiːnɒl])	五氯苯酚
TeCP (2, 3, 5, 6-Tetrachlorophenol [ˌtetrəˌklɒrəˈfiːnɒl])	2, 3, 5, 6-四氯苯酚
mould [məʊld]	发霉
fungi [ˈfʌŋgaɪ]	真菌
chlorinated organic carrier	含氯的有机染色载体
PCB (polychlorinated [ˌpɒlɪˈklɔːrɪneɪtɪd] biphenyl [baɪˈfiːnɪl])	多氯联苯
PCT (polychlorinated terphenyl [təˈfiːnɪl])	多氯三联苯
softener [ˈsɒfnə]	柔软剂
carrier	染色载体
flame retardant [rɪˈtɑːdənt]	阻燃剂
liver [ˈlɪvə]	肝脏
hormone [ˈhɔːməʊn]	荷尔蒙, 激素
immune [iˈmjuːn]	免疫的
nervous [ˈnɜːvəs]	神经的
Tributyltin [traɪˈbjuːtɪltɪn](TBT)	三丁基锡
anti-microbial [ˈæntɪ-maɪˈkrəʊbɪəl] finishing	抗微生物整理
concentration [ˌkɒnsənˈtreɪʃən]	浓度
mother of pearl	贝母
mechanical strength	机械强度
bursting strength	顶破强度
tearing strength	撕裂强度
fabric fault	织物疵点
colour spot	色花
oil spot	油污迹
hole	破洞

streak	条痕
significant level	显著性等级
skipped stitch	跳线
thread breakage	断线
irregular stitch	线迹不匀
damage to fabric	织物损坏
seam pucker	缝迹起皱
thread tension	缝线张力
burrs on the needle	毛针
stay tape	接缝狭带，牵条
ironing mark	烫痕
glazed or yellowish spots	极光或焦斑
overpressing	过分熨烫
wrinkle	褶皱

Exercises

A. Please use "T" or "F" to indicate whether the following statements are TRUE or FALSE：

(1) Garment quality issues are closely related to the production cost.　　　　(　　)

(2) There should be some tolerance for garment quality specifications.　　　(　　)

(3) All garments to be exported must follow the standards of the buyer's country.　(　　)

(4) If there are no detailed contractual terms regarding the rainproof quality of the fabric, the buyer cannot complain if the rain coats under that contract are not rainproof.　(　　)

(5) Garment inherent quality is, to a great extent, related to the quality of fabrics.　(　　)

(6) When classifying colour fastness, the higher the grade number, the poorer is the colour fastness.　(　　)

(7) Poor management in garment making-up could lead to hazards for wearers.　(　　)

(8) It is unlikely that a claim will be entertained by the fabric manufacturer once the fabric has been cut.　(　　)

(9) The word "overpressing" means "top pressing" in garment manufacture.　(　　)

(10) Any colour shading on fabrics is not allowed because it will cause subsequent colour shading on garments.　(　　)

B. Please select the word(s) or phrase(s) that makes the statement correct：

(1) The supplier should ensure that the quality of the garments to be exported _____.　(　　)

    a) must comply with the contractual terms

    b) should conform to the relative governmental regulations

    c) should be fit for the particular purpose expressly or impliedly made known to the seller when the contract was concluded.

    d) should always be the same as in previous orders

(2) Which of the followings may be considered as an issue associated with an inherent garment quality?　　　　　　　　　　　　　　　　　　　　　　　　　　　（　　）

　　a) Poor colour fastness

　　b) Insufficient mechanical strength

　　c) Seam puckering

　　d) Residual pesticides

(3) Which of the followings may cause garment apparent quality problems?　　　（　　）

　　a) Poor sewing skill

　　b) Incorrect setting of the ironing temperature

　　c) Azo dyes used

　　d) Non-durable buttons of mother of pearl used on children's garments

(4) Which of the following locations are referred to as the garment parts that fall within the most significant quality level?　　　　　　　　　　　　　　　　　　　（　　）

　　a) The lower part of a front facing　　　b) Top collar

　　c) Top of the shirt yoke　　　　　　　　d) Under sleeve

(5) Seam puckering may be caused because of _____.　　　　　　　　（　　）

　　a) excessive sewing thread tensions

　　b) poor operator skill

　　c) burrs on the needle

　　d) distortion of fabric structure

Reading Materials

Tolerable range of abdomen and waist skin temperature
for heating-capable smart garments

By Soyoung Kim, Kyunghi Hong, Heeran Lee

Abstract

Purpose: This study aims to provide information on how to monitor the temperature setting of a heating device in order to implement a heating unit successfully in the smart clothing by observing voluntary heating behavior of wearers.

Design/ methodology/ approach: Subjects wearing base layers and additional clothing were asked to turn on and off the switch when wanted in the cold environmental chamber. Tolerable range of skin temperature (TST) depending on the location of body was obtained by observing the temperature at the time when the heating device was turned on and off during a rest-running-rest protocol.

Findings: The TST was 32.8−49.4 ℃ and decreased to 31.3−37.6 ℃ around abdomen and back waist, respectively. Changes in the wearers' voluntary control behavior were observed depending on the individual's level of cold-sensitivity and activity level of rest and running. TST was 35.8−49.4 ℃ (Rest 1: rest before exercise), 40.0−42.0 ℃ (Running) and 35.3−

43.2 ℃（Rest 2：rest after exercise）for cold-sensitive group，whereas it was 32.8–36.2 ℃（Running）and 34.4–45.7 ℃（Rest 2：rest after exercise）for cold-insensitive group.

Originality/ Value：In this study，results with detailed body locations and wearer's thermal sensitivity provide practical references for the implementation of a heating device to the comfortable multilayered smart clothing.

—excerpted from *International Journal of Clothing Science and Technology*，Vol. 33，No. 6 pages 929–941.

【参考提示】

1. *International Journal of Clothing Science and Technology* 是英国 Emerald 出版公司出版的关于服装科技方面的学术刊物（www. emerald. com/insight/publication/issn/0955-6222）。

2. Subjects 参与实验的对象。

Factors Influencing Consumers' Intention to Adopt Fashion Robot Advisors：Psychological Network Analysis

So Young Song，Youn-Kyung Kim

Abstract

Drawing upon the theory of human-robot interaction（HRI），this study examined the relations among perceived characteristics of fashion robot advisors（FRAs），consumers' negative preconceptions toward robots，and positive dispositions toward technology to identify network differences in adoption and nonadoption groups. For interviews，pretests，and main data collection，we presented video clips of FRAs as stimuli. Based on the data（n＝464）collected via an online survey，we conducted psychological network analysis to explore defining factors that differentiate adoption and nonadoption groups. The results indicate that perceived characteristics of social intelligence，human-likeness，and knowledgeableness combined with a positive disposition of technological self-efficacy lead to adoption of FRAs. This study contributes to the literature on the theory of HRI and technology acceptance models，particularly in fashion retail sectors. Furthermore，this study provides a new graphical approach to networks that conceptualizes shoppers' adoption of technology as a complex interplay of psychological attributes.

—excerpted from *Clothing and Textiles Research Journal*，Volume 40，Issue 1，January 2022，page 3.

【参考提示】

1. *Clothing and Textiles Research Journal* 是位于美国的国际纺织服装协会（International Textile and Apparel Association，www. itaaonline . org）的关于纺织和服装专业方面研究的学术性杂志，网页版由 SAGE 公司发布（https://journals. sagepub. com/home/ctr）。

2. fashion robot advisor 机器人（智能）时装顾问。

3. video clips 视频剪辑。

4. human-likeness 与人类相似度。

5. Knowledgeableness 博识度。

6. self-efficacy 自我效能感。psychological attribute 心理属性。

Quality matters: reviewing the connections between perceived quality and clothing use time

Maarit Aakko and Kirsi Niinimäki

School of Arts, Design and Architecture, Aalto University, Espoo, Finland

Abstract

Purpose —Extending the active lifetimes of garments by producing better quality is a widely discussed strategy for reducing environmental impacts of the garment industry. While quality is an important aspect of clothing, the concept of quality is ambiguous, and, moreover, consumers may perceive quality in individual ways. Therefore, it is important to deepen the general understanding regarding the quality of clothing.

Design/ methodology/ approach —This paper presents an integrated literature review of the recent discussion of perceived quality of clothing and of the links between quality and clothing lifetimes; 47 selected articles and other literature obtained primarily through fashion/ clothing/ apparel journals were included in this review.

Findings —The main ideas from the articles are thematized into the following sections: the process of assessment, levels involved in assessment, multidimensional cues of assessment, and quality and clothing use times. The paper highlights that perceiving quality is a process guided by both expectations and experience, and assembles the various aspects into a conceptual map that depicts the connections between the conceptual levels involved in assessing quality. It also illustrates connections between quality and clothing use times.

Research limitations/ implications —This paper focused on perceived quality on a conceptual level. Further studies could examine and establish deeper links between quality, sustainability and garment lifespans.

Originality/ value—The study draws together studies on perceived quality, presenting the foundational literature and key concepts of quality of clothing. It summarizes them in a conceptual map that may help visualize various aspects affecting the assessment of quality and deepen the general understanding of the quality of garments.

—excerpted from *Journal of Fashion Marketing and Management*, Vol. 26, No. 1, 2022, page 107.

【参考提示】

1. *Journal of Fashion Marketing and Management* 也是 Emerald 出版公司下的关于时装营销及管理方面的学术刊物(https://www. emeraldgrouppublishing. com/journal/jfmm)。

2. conceptual map 概念图。

Keys to Exercises

Chapter 1
A. (1) T; (2) T; (3) F; (4) T; (5) T; (6) T; (7) F; 8) T; (9) T; (10) T
B. (1) d; (2) a, c, d; (3) b, d; (4) a, b, c, d; (5) a; (6) c

Chapter 2
A. (1) T; (2) T; (3) T; (4) T; (5) F; (6) F; (7) F; (8) F; (9) T; (10) T
B. (1) c, d; (2) a, c; (3) b, c; (4) a, b, c, d; (5) d
C. left front, right front, left front side, right front side, left front facing, right front facing, left back, right back, left back side, right back side, back neck facing, left top collar, right top collar, left under collar, right under collar

Chapter 3
A. (1) T; (2) T; (3) F; (4) T; (5) F; (6) F; (7) F; (8) F; (9) T; (10) F
B. (1) a, c, d; (2) b; (3) b, c; (4) b, c, d; (5) a, d; (6) a, b, c

Chapter 4
A. (1) T; (2) F; (3) F; (4) F; (5) T; (6) F; (7) T; (8) F; (9) F; (10) F
B. (1) c; (2) a; (3) d; (4) a; (5) b, c, d

Chapter 5
A. (1) T; (2) F; (3) T; (4) T; (5) F; (6) T; (7) T; (8) T; (9) T; (10) F
B. (1) a, b, c; (2) a, b, c, d; (3) b, c; (4) a, b, c, d; (5) a

Chapter 6
A. (1) T; (2) T; (3) F; (4) F; (5) T; (6) F; (7) T; (8) T; (9) F; (10) F
B. (1) a, b, c; (2) a, b, d; (3) a, b; (4) b, c; (5) a, b, d

Chinese Version

第 一 章　服 装 分 类

1　日常分类

　　服装可有数种分类方法。例如,根据所穿着的"层次",日常服装可以分为内衣和外衣。内衣是紧贴身体穿着的衣服,即,穿着在其他衣服下的服装。外衣,顾名思义,指的是穿在内衣外的衣服。然而,如今要定义一件衣服是内衣还是外衣可能会有难度。比如,吊带背心可能被女孩子穿作内衣或夏季的"外衣"。短袖针织圆领衫历史上被视作内衣,如今在公共场合穿着也已广泛被认可。

　　另外一种常用的服装分类法是根据服装所着的半身位置,即,从颈部向下至所谓腰部的上半身或从腰部向下到双脚的下半身,将一件服装分为上装或下装。应该注意的是,一件覆盖身体大多数部位的服装中,上面部分可称为"garment top"而不是"top garment",下面部分可称为"garment bottom"而不是"bottom garment"。术语"top garment"和"bottom garment"还会造成混淆,因为"top garment"也可以指外衣,"bottom garment"也可以指内衣。此外,可以仅用"top"或"bottoms"本身来指上装或下装。

1.1　上装

　　如果从外到内或从上到下观察一套衣服,有可能看到通常所穿的外衣上装是大衣(风衣,coat)。(英语中)"coat"这词实际是一个含义非常广的词(统称)用于命名一件通常为长袖并且长度到从膝上某个位置至脚踝的衣服。根据它们的款式、涉及的布料以及它们的用途等,有(厚)大衣、(花呢)轻便大衣,防尘大衣(风衣),军旅式雨衣,休闲大衣,雨衣,(全天候)风雨衣等。有时描述这类服装时,(英语中)不一定必须使用"coat"这个词,比如"parka"(派克大衣)就是一类休闲大衣。

　　"jacket"也是一类上装外套,它可以穿在大衣内。广义地说,"jacket"是几乎所有短上装外衣的统称。一件"jacket"可以是一款正式的或休闲的服装。作为正装,它是度身定做的服装,单排纽或双排纽,背后开衩或不开衩,如果开衩,或是单衩或是双衩,并且为驳领。它们可以作为(西服)套装的一部分穿着,通常和相匹配的裤装一起;也可以作为单独的服装穿着,再配更正式的或更休闲的裤子。正式的无燕尾的晚礼服上装在正式场合穿,而普通的或日常的西服套装在不很正式的场合或办公室穿。"blazer"是带驳领的上装短外套,通常为双排纽的,虽然它传统上和制服有关,比如航空乘务员,学生以及游艇俱乐部成员等穿的制服,但现在越来越多地被普通大众作为外衣上装穿着。狭义地说,一件休闲的"jacket"是长袖休闲上装短外套(即夹克衫),通常装有腰头或可用绳收腰。"blouson"就是一种宽松的夹克,通常装大袖窿并且装宽紧带的腰。在贸易中,夹克式样并且衣长相对较长的服装会归类于休闲大衣。诸如防风上衣,带风帽的风衣和风帽开口较小的派克衫等通常归属风衣或夹克的类别,因此,人们倾向于将衣长作为确定服装归属类别的主要因素之一。

　　针织毛衣是针织服装,通常可以单独穿或穿在外套上装或大衣内。不考虑它的花型设计或线圈结构,针织毛衣可以分为套衫(因为它是通过套上头往下拉而穿上身的)和开衫(英语中开衫为"cardigan",该词也用于带集圈横列的一类针织结构)。开衫前部用纽扣从上往下扣合。在英国英语中,套衫又被称为"jumper",相对照的是,在美国英语中,"jumper"可能指无袖连衣裙,通常穿在女式衬衣外。它或被称为"jumper dress",而在英国英语中,被称为"pinafore dress"。

　　称为"shirt"的衬衫可以男穿和女穿,可以作为一款办公室穿的正式服装,常配领带。在美国,这样的衬衫会被称为"dress shirt"。传统上,正式的衬衫由机织面料制成,带有领子、覆肩和袖口。典型的衬衫领由领子和领脚组成,上有衬头以提供硬挺的领子形状,但如今立领衬衫到处可见,尤其在一些商务办公楼里。与称为"shirt"的衬衫相比,称为"blouse"的女式衬衫通常由女士穿着(但在美国,"blouse"这个词也指一款男式军用制服)。女式衬衫可以是长袖的或短袖的,坦领或领带领,或甚至无领有时还无袖。领或袖口加上花边特色可赋予服装更强的女性化气质。女式衬衫可以有很多袖或领的款式变化,而称为"shirt"的衬衫,它的款式,尤其是袖子,或多或少总是一成不变的:短袖或带有袖衩和扣纽扣的袖口的长袖。应该指出,英语中服装名称含有"shirt"的并不总是指上面所提及的那种款式的衬衫,"T-shirt"和"sweatshirt"可能就是很好的例子。它们通常是针织圆领的,无领子和衣袋。"sweatshirt"通常用针织起绒布制成,长袖;"T-shirt"通常为短袖。如今,连风帽的拉链"sweatshirt"在年轻人中很受欢迎。"polo shirt"通常是棉制的,休闲短袖针织衫,前部有带纽扣的开口,开口的大小足以使它能够套上头并拉下。没有袖子的"T-shirt"可能被称为"tank top"。

"dress"这个词可以为可数名词或不可数名词。作为可数名词,该词仅用于女式服装。"dress"是个统称的词,任何带有上身成为一体的裙子都能够被称为"dress",即连衣裙。连衣裙有很多的种类,诸如称为"pinafore dress"的无袖连衣裙(如前所述,美国英语中被称为"jumper dress")和露肩背带式连衣裙等。作为不可数名词,"dress"有时和其他词一起形成复合词,以表示某专类的服装,比如称为"national dress"的"民族服装";称为"evening dress"的"晚礼服"和称为"working dress"的"工作服"。在这种情况下,这类表达可能用于男装或女装。(英语中)"to dress oneself"指"穿上衣服"。

最常见的内衣上装为背心,即"vest"或"singlet"。按定义,"vest"通常指的是无袖上衣。然而,它不一定指内衣,比如,称为"waistcoat"的西装马夹(在美国)被称为"vest"。根据它的一般含义,"vest"是一类内衣上装,因此,除了有无袖的"vest",也可以有长袖的或短袖的"vest"。

对于女士,称为"slip"的衬裙是一类内衣上装(当然,可能有称为"half slip"的下装衬裙)。另外,还有女士的贴身内衣,诸如文胸、束腰塑身内衣和胸衣等,所有这些都归类于女内衣。

1.2 下装

最常见的下装外套是长裤。在美国英语中,使用术语"pants"或"slacks"而不是"trousers"。然而,"pants"在英国英语中指的是内衣下装(可能是"underpants"的缩写,即,穿在"pants"内(under)的),而内衣下装在美国英语中被称为"underpants"。

术语"长裤"是一个包含一切的术语,其区别在于正装裤子是西服套装的一部分或去办公时穿;工装裤在较多体力劳动的场合穿;休闲裤是悠闲时的着装。如今还出现一些术语,如"chinos"用于描述一种款式的休闲裤,当然,还有"jeans",即牛仔裤。称为"slacks"的宽松裤通常是描述女式和男式(在美国)休闲裤的术语,它常常(但不总是)为了舒适在腰部装有宽紧带。

长裤通常从腰延伸到双脚,到膝盖上下位置的被称为短裤。游泳穿的短裤称为"trunks",即游泳裤(swimming trunks)。马裤、牛仔裤和带护胸的背带工装裤等都是常见的裤子的款式。可以改变裤的外形轮廓来获得不同款式的裤子,其中,小裤腿、喇叭裤或直筒裤等都是典型的例子。通常,百慕大式短裤和一些裤子在裤脚边有翻边。其他在裤袋或裤腰的变化也能够产生新的裤子款式。宽松裤和脚踝处开衩并用拉链关闭的裤子都是如今非常流行的休闲外衣下装,如果用作运动装,它们可能被称为"慢跑裤"。带"猫须"的石洗牛仔裤也非常受青年人欢迎。

裙子是女士的下装,然而,男人穿的苏格兰短裙也是一种裙子,传统上由苏格兰男性士兵穿着,如今似乎只有在某些传统仪式上或体育集会上穿,以强调对苏格兰的爱国情操。此外,纱丽和纱笼也是一种裹紧打结或塞紧的裙子,常常由热带国家土生土长的妇女或旅游者穿。在那些国家,男人同样也穿相似款式的服装,不过人们倾向于用本民族的名称而不是莎笼来称呼它们。

裙子也有多种款式,比如A字裙、直筒裙、(上宽下窄的)陀螺裙、(打裥的)百褶裙、拼片裙、超短裙和围裹裙等等。显然,改变长度和外形轮廓,或加入一些款式细节,可以设计出相当多种类的裙子。裙子款式变化的另外的例子有加拼腰的裙子、高腰或低腰的裙子以及无腰头的裙子等等。宽裤脚的短裤被称为裙裤,即"culottes"或"divided skirts"。

正如以上所提及的,"pants"或"underpants"是广泛使用的描述内衣下装的词。三角裤是非常常见的用于称呼男士和女士穿的内衣下装的术语,然而,除了三角裤,对女孩或女士来说"pants"、"panties"或"knickers"是更常用的术语。虽然从技术上说只有较长的"三角裤"能够被称为"knickers"(英国英语,或美国英语中为"panties"),术语"knickers"在英国可以和"pants"替换使用。应该注意的是,"knickers"这个词在美国也可作为"knickerbockers"(即短灯笼裤)的缩写,短灯笼裤是有点老式的膝部鼓起的男人或男孩穿的裤子。布带裤和比基尼裤指的是女性穿的较暴露的内衣,后者由两件套的比基尼泳装派生而来。男式内衣通过款式特征定义,比如称为"Y-fronts"的前开裆短内裤和称为"boxer shorts"的平脚裤。称为"tights"的紧身裤和称为"panty-hose"的连裤袜只有妇女穿着。

1.3 西服套装和便服套装

上装和下装总是分开销售。妇女通常喜欢购买上下装分开销售的服装,通过细心选择搭配上下装,她们能够创造出通常反映自己个性的各种时尚风格。

通常由相同面料制成并具备相同的款式特征的两件或三件衣服构成的一套服装将可能被称为"套装"。一套西服套装通常由一件上装和一条长裤(或,对于女套装,一件上装和一条裙子)组成。三件套西服套装包括一件称为"waistcoat"的西装马夹(如前文所提及,在美国还称它为"vest")。一般而言,男式或女式的两件套或三件套通常在正式场合穿着或作为办公着装。作为传统的标准西方男士商务装,西服套装通常和长袖衬衫和领带一起穿戴。在非常正式场合穿的更正式的着装是正餐礼服套装或无燕尾的晚礼服,或,有时在接待仪式上穿着的或由婚礼上的新郎穿的常礼服或婚礼服套装。然而,有些其他的"套装",如运动套装、针织套装等,是比较休闲的套装。如果服装(不指连衣裙)被设计成上下装连成一体的,"suit"这个词也可能会用来构成复合词来描述该类服装,比如称为"jumpsuit"的连裤外衣和称为"body suit"的连短裤内衣等。"ensemble"或"set"是"suit"之外的类似的词,可能用来描述上下装在某些方面相协调的一套服装。

除了上述的分类,还有其他的一些方法对服装分类或命名。某些服装名称从它们最初的用途派生而来,比如,运动服、网球衫、防尘服、沙滩装和裙摆及地的女装晚礼服等;有的服装名称来自于它们原创者的名字或随它们流行的地方名称而定,比如被称为"Mackintosh"的雨衣(发明者名字为"Charles Macintosh")、阿拉伯长袍和夏威夷衬衫等。在国际贸易中,通常有必要说明服装是男装、女装或童装,比如,男风衣、女风衣或童风衣。如何命名服装,或多或少源于传统或人们的偏好。

2 国际贸易中的分类

服装贸易,特别是发达国家与发展中国家间的服装贸易,是国际贸易中的重要领域之一。为了监控服装贸易,人们已对服装建立了正式公认的分类方法。这使得各自国家的海关税务部门能够编制有关服装进出口的统计资料。

2.1 基于协调制编码的分类

最好的服装分类方法之一是采用协调制商品名称和编码方法,简称为协调制编码。

在协调制商品名称和编码方法中,纺织品和纺织物属第11类。针织或钩编的服装一般归类于第61章中;非针织或钩编的服装,或不是用针织面料制作的服装在第62章中。某些其他章目也与服装有关,比如,旧衣物属于第63章,毛皮服装和带毛皮夹里的服装属第43章。

首先必须关注每一章的注释,它对于每章的使用方法给出专门的解释。比如,"suit"这个词在61章和62章的注释中被严格定义。根据这些注释,术语"西服套装"指一套由两件或三件组成的服装,用相同的面料制成,包含:

☐ "一件设计成由上半身穿着的外套大衣或短上装,除袖子外,构成它的面料的衣片由四片或四片以上组成,可能还带一件度身定制的西装马夹,马夹的前片面料和该套服装其他各件的面料相同,马夹的后片和外套大衣或短上装的里料相同。以及

☐ 一件设计成由下半身穿着的服装,不可有背带或护胸,可以是长裤、马裤或短裤(游泳装除外)、裙

子或裙裤。"

此外,注释强调,"西服套装"各件衣服必须具备相同的织物结构、相同的颜色和相同的成分;它们必须是相同的款式和相应或般配的尺码,不过,它们可以使用不同织物的滚边。

根据这些注释,术语"suits"也包括如下服装套装,不论它们是否满足上述条件:

☐ 常礼服,包括一件素色的短上装,下垂的圆弧后摆(下摆为圆角的),和带条纹的长裤;

☐ 晚礼服(燕尾服),一般用黑色织物制成,上装前部相对较短,不闭合且臀部处裁成狭窄的下垂燕尾;

☐ 没有燕尾的晚礼服套装,其中上装款式和普通短上装相似(虽然或许露出更多的衬衫前部)但有光滑的丝绸或仿丝绸的驳领。

其他一些术语,比如"便服套装"和"成套服装"在第61章和62章的注释中也已严格定义。

2.2 基于特殊规则的分类

为了监控服装进口,某些国家或区域制定了自己的服装分类方法。这些方法间可能有很大的差异。在欧盟和加拿大的方法中,服装根据它们的类型和构成的织物分类,比如,机织物的长裤在欧盟类别下归为6类,加拿大类别下归为5类。

在美国,分类方法完全不同,它使用了三位数的类别号。第一位数字表示服装使用的纺织材料(比如,2为棉及/或化纤制的服装;3为棉制服装;4为毛制服装;6为化纤制服装;8为真丝混纺或非棉植物性纤维制服装)。另外两位数字表示服装的类型。比如,棉制男式长裤归在类别号347下,棉制女式长裤归在类别号348下。如果同样的长裤是毛制的,它们将归入类别号447和448下。

2.3 基于生产方法的分类

在国际贸易中,区分原产国非常重要,因为进口国的海关税务部门根据原产国和政府的贸易政策来决定征收何种关税。如果一件服装的所有制作过程发生在一个国家,确定原产国很简单,但是,如果涉及进口材料,或服装在不同的国家制作,如何区分原产国的规则变得很重要。

各国执行不同的规则以确定服装的原产国,但多数国家根据服装生产"主工序"发生地制定确定原产地的规则。然而,何为"主工序"似乎是一个有争议的问题。许多国家可能会将服装分成裁剪成形和编织成形的。对于裁剪成形的服装,裁剪后的衣片缝合所在国为原产国;对于编织成形的服装,服装或衣片(通过放针或收针)编织所在国定义为原产国。必须注意,一个国家非常有可能根据服装进口的情况调整它的原产地标准。将原产国规则标准化是世界贸易组织非常期盼的事。

3 分类的重要性

日常生活中服装分类的方法不很重要,然而,在贸易中尤其是在国际贸易中,由于以下原因,有必要正确地对服装进行分类和命名。

3.1 为了统计

每一个国家的有关当局必须编制进出口统计数据,这样,根据这些统计数据,政府可以调整进出口政策。然而,没有对服装科学的分类方法,不可能有效地得出服装贸易的统计数据。无论是协调制编码、欧盟分类方法或美国分类方法,它们想提供公认的分类方法的意图是相同的。

政府或区域的原产国规则会影响统计。比如,根据美国协调关税制,由美国生产的织物制成的服装(裁剪并缝合)有资格在纺织品服装外加工安排下入关,然而,这样的资格必须符合纺织品协议执行委员会制定的程序。进口商被要求在入境报关总表上恰当的十位数协调关税编码上以符号"S"作为前缀以区别那些服装。

3.2 为了描述合同中的商品名称

服装的名称是服装进口或出口合同中商品描述的一部分。因此,如果人们知道服装如何分类,他们可能就会知道如何为这类合同命名服装。然而,在合同中落实或接受服装名称时,仍应对它加以关注。

首先,相同的或相似的服装名称在不同的地区或国家可能会有不同的含义。如前所述,"pants"这个词在美国英语中指的是外套下装,但在英国英语中通常指的是内衣下装。所谓的"T-shirt"在中国通常指一种有领的针织衬衫而不是更常见的定义,即,短袖无领子的圆领针织衫,它在摊平时,像字母"T"。针织衬衫或有领的针织衫应该在协调制编码中相应归类于 H.S.6105 或 H.S.6106,而"T-shirt",无领针织衫,应归属 H.S.6109。

其次,在国际贸易中,英语是商务单证中广泛使用的语言,因此,使用英语为母语者通常使用一些措词来描述合同中的服装,这一点非常重要。任何根据母语为非英语的人想象或派生出来的新的表述都有可能造成混淆,并且可能导致潜在的日后贸易争议。当卖方向他们潜在的客户寄价目表或买方寄出他们要求报价的询盘时也应该关注这个问题。

第三,如果所涉及的材料包含在服装的描述中,应该给出正确材料成分。不过,协调制编码中,两种或两种以上纺织材料混纺的织物归类应该视其好像完全由其中比重大的纺织材料制成,而不是按各单一的纺织材料归类。比如,85%的毛和15%的腈纶混纺纱制成的针织套衫将作为毛的套衫归类为 H.S.6110.11。但是合同中必须给出正确的混纺比。绝对不宜在合同中将该服装仅描述为"毛套衫"而是更精确地描述为"套衫,85%毛/15%腈纶"。

最后,如可能,在进口或出口合同中的服装名称应该和协调制编码中的名称一致;或如果贸易在某些政府双边协议中受限,应该和相应协议类别中的一致。

如果交易涉及多款服装,合同可以给出服装统称,然后说明对面料、里料和辅料的质量要求,并且列出款号和相应的数量及单价。当然,无论在合同中如何命名服装,可以使用样品和工艺单确切阐明所涉及的产品。为合同顺利履行,相互理解是必要的。

图 1.1　大衣
1—双排纽大衣　2—防尘大衣　3—军旅式雨衣
4—薄大衣　5— 派克大衣

图 1.2　短上装
1—宽松式夹克　2—西便装　3—带风帽的短风衣
4—防风上衣　5—风帽开口很小的派克衫

图 1.3　与"shirt"有关的服装
1—长袖运动衫和短裤　2—短袖针织圆领衫和带背带护胸的工装裤
3—宽松式衬衫和牛仔裤　4—连裤外衣　5—短袖圆领衫　6、7—衬衫（正装衬衫）
1—长袖圆领运动衫和短裤　2—短袖圆领衫和带护胸的背带工装裤

图 1.4　称为"blouses"的服装
1—领带领女式衬衫　2—无袖女衬衫　3—方领圈女衬衫
4—短袖女衬衫　5—"透视"装女衬衫　6—（美）军上装

图 1.5　连衣裙和裙子
1—无袖连衣裙　2—露肩背带式连衣裙　3—夏威夷连衣裙
4—无袖连衣裙　5—套衫和百褶裙　6—裙套

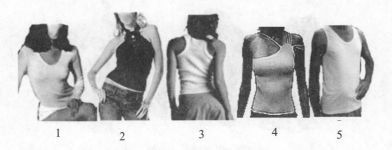

图 1.6　称为"vest"的服装和背心
1—长袖内衣　2—吊颈背心　3—窄背背心　4—多带式吊带衫　5—男式背心

图 1.7　内衣
1—胸衣和三角裤　2—束腰塑身内衣、女内裤和长袜　3 和 4—连短裤内衣
5—平脚裤　6—（男）前开裆短内裤　7—布带裤　8—女短裤

图 1.8　裙
1—A 字裙　2—直筒裙　3—拼片裙　4—褶裙　5—网球围裹裙

图 1.9　女晚礼服、套装等
1—裙摆及地的女晚礼服　2—运动套衫　3—西装马夹
4—燕尾式大衣　5—正餐礼服套装

图 1.10　纱笼、纱丽等
1—纱丽　2—纱笼　3—沙滩装　4—沙滩短裤　5—泳衣

第二章 服装组成部分、样板和测量部位

1 服装组成部分

一件服装可能有数个部分,这取决于它的款式特征。各个部分可能由一片或更多的相同或不同织物的衣片组成,衣片通过缝纫、热熔黏合、胶水黏合或焊接拼合在一起构成一件服装。

1.1 上装的组成部分

最简单的上装可能只有大身。换言之,它是无袖和无领的。与此相反的是,相对复杂的款式可能有许多的款式特征。

1.1.1 大身

大多数服装的大身有单独的由摆缝缝合的前片和后片。某些针织服装(比如某些短袖针织圆领衫)是圆筒式编织的,没有摆缝。此外,如果服装在前部完全分开,它将有左前片和右前片。与此相似,如果对称,后部可以由背中缝缝合的左后片和右后片组成。某些服装用前挂面和后领圈加固(比如某些女式衬衫或衬衫领的部位),因此,如需要,左前挂面、右前挂面和后领圈的衣片可能是必不可少的。某些女式上装的款式特征为公主缝。如果有摆缝并前后部都有公主缝,就需要左后侧片、右后侧片、左前侧片和右前侧片的衣片(见图 2.1)

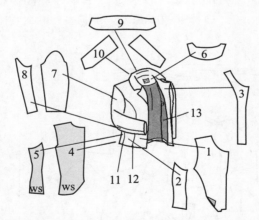

图 2.1 显示某些面料衣片的典型正装短上衣示意图(注:"WS"指"反面")
1—右前片 2—右前侧片 3—左前挂面 4—右后片 5—右后侧片 6—后领圈
7—右大袖 8—右小袖 9—面领 10—右底领 11—摆缝 12—公主缝 13—夹里

设计师有时为了款式变化而使用分割缝。这样的缝子把有关衣片分解成独立的衣片。比如,如果分割缝横过大身前部出现,其结果会使前片分解为上前片和下前片(见图 2.2)。

1.1.2 袖

许多服装上一片袖,因此,只需左袖和右袖两块衣片。对于袖山头相对较高的袖子通常采用带大袖和小袖的两片袖,在这种情况下应该有左大袖、左小袖、右大袖及右小袖的衣片。插肩袖可以由前袖片和后袖片组成(见图 2.2)。

典型的衬衫袖为带袖口的一片袖。衬衫袖口的衣片数可以为两片或四片,在前者情况下缝合时,每一袖口衣片将正面朝里对折,左右两端缝合,然后将里面外翻(称为"bagged out",即像翻袋子那样被翻转)。在后者情况下,一个袖口的两片袖口衣片面对面放在一起,将两端和顶端缝合(见图 2.3)。某

些设计师情愿使用后一种方法,因为袖口衣片尺寸较小,这有助于排料时增加织物的利用率(排料是将样板在织物上配置以待裁剪的工序,它的描述可参见第五章1.1节)。某些夹克衫的袖口只是由袖口边翻折而成,这样,无需单独的袖口衣片。

多数袖子属装袖,它们缝合在服装袖窿上。不过,袖子也可以设计成服装大身衣片相连的一部分。

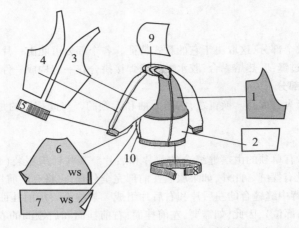

图2.2　显示某些衣片的典型宽松式夹克衫的示意图
1—左上前片　2—左下前片　3—右前袖(插肩袖)　4—右后袖(插肩袖)
5—针织罗纹袖口　6—上后片　7—下后片　8—针织罗纹腰头
9—右风帽　10—摆缝

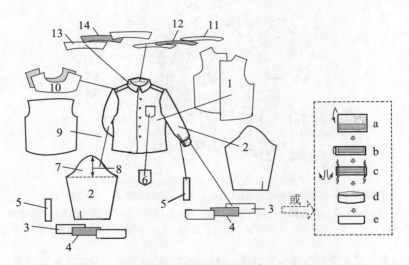

图2.3　普通衬衫的衣片
1—前片　2—袖片　3—袖口　4—袖口衬头　5—袖衩滚条　6—贴袋　7—袖山
8—袖山高　9—后片　10—覆肩　11—领脚　12—领脚衬头　13—领子　14—领衬
a)反面上衬　b)正面朝内折叠　c)缝合两端　d)正面朝外翻　e)袖口在小烫后备妥

设计服装时,根据所需的款式特征可以确定各种袖长。男式上装被设计成或为长袖或为短袖,与此相反,女式服装除长袖或短袖外,可能还有帽形袖、中袖、3/4袖。袖的款式变化有许多,比如插肩袖、多尔曼袖、蝙蝠袖、钟形袖和主教袖等。此外,泡泡袖、瓜形袖、披肩袖、郁金花式袖都是流行的连衣裙短袖(见图2.4)。

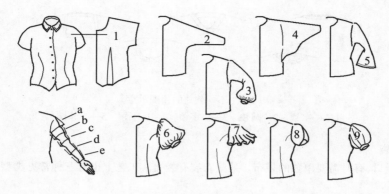

图 2.4　一些袖子款式变化示意
1—连身袖　2—多尔曼袖　3—主教袖　4—蝙蝠袖
5—钟形袖　6—泡泡袖　7—披肩短袖　8—郁金花式短袖　9—瓜形袖
a)帽形袖　b)短袖　c)中袖　d)3/4 袖　e)长袖

1.1.3　领子

领子有三种基本类型。最常见的是装上去的衣领,它又可再分为坦领和立领。对于这样的领子,须有面领和底领衣片。伊顿领是一种典型的坦领,中式领是典型的立领。衬衫领实际是坦领和立领的复合,它包含四片衣片,即,面领、底领、面领脚和底领脚(见图 2.3)。

连身领是另一种类型。对于连身领,底领通常是大身衣片的一部分,面领是前挂面的一部分,这样,没有单独的领子衣片。翻领和披肩领是连身领很好的例子。

具有符合第三种基本类型领子的服装带有衣领和驳领。衣领可以由独立的面领和底领衣片制成,底领可以再分成斜向丝缕线的左、右部分。驳领的底下部分是前片的一部分,驳领的面子部分是前挂面的一部分(见图 2.5)。

领高和领子形状的变化带来各种领子的款式。对于无领子的服装,通过变化服装领圈获得款式变化。一字领、圆领、V 字领、U 字领和方领圈是很常见的领圈款式(见图 2.6)。某些无领运动衫带有直接装在领圈上的风帽,某些夹克衫在衣领后带有可脱卸式风帽。

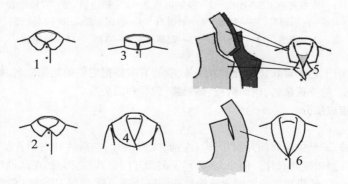

图 2.5　不同款式领子的示意图
1—彼得潘领(坦领)　2—伊顿领　3—立领　4—披肩领
5—带驳领的衣领　6—翻领(连身领)

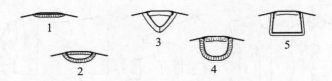

图 2.6 不同款式的领圈示意
1——一字领 2—圆领 3—V 字领 4—U 字领 5—方领圈

1.1.4 衣袋

衣袋需要袋口,有时需要单独的袋布。某些衣袋有袋盖。工艺上也有三种常见类型的衣袋:挖袋、贴袋和结构袋。

挖袋的袋口是在衣片中按设计装衣袋的地方剪出的。最常见的挖袋款式有嵌线袋和贴边袋。制作袋口还需要嵌线滚边或袋贴边的衣片。

结构袋的袋口由衣片边缘形成,袋口勿需额外的衣片。骑缝袋是结构袋的一个常见的例子。

就成衣工艺而言,贴袋可能是最简单的衣袋。衬衫衣袋是一种贴袋,风箱袋也是(见图 2.7)。由于衣袋由衣袋袋布和衣袋所贴的衣片形成,不需单独的袋布衣片。然而,如果使用的织物有条子或方格花型,通常在成衣时有必要将衣袋上的条子或方格与底下衣片上的花型对齐。因为花型错位很惹人注目,另外的做法是(特别是方格花型时),小心地将衣袋按方格花型成角度裁剪,这样当它被缝到服装衣片上时,它和方格花型形成明显的反差。

图 2.7 衣袋、搭襻和"yoke"等
1—大衣过肩 2—袖襻 3—拉链前直条 4—贴边袋 5—双嵌线袋
6—衬衫覆肩 7—背襻 8—袖拼肩 9—领圈滚条 10—裙拼腰
11—袋垫 12—结构袋 13—蚂蟥襻 14—袋盖 15—风箱袋

除了以上提及的工艺上对衣袋的分类方法外,衣袋有时根据它们的功能命名,如零钱袋或表袋等;或根据它们在服装上的位置命名,比如胸袋、裤腿袋、臀袋和内袋。

1.2 下装的组成部分

1.2.1 裤

通常组成一条裤子至少需要四片衣片,即,左前片、右前片、左后片和右后片。如果要上腰头并且要有裤袋的话,还需要相应的衣片。而今,多数裤子前部开门襟,谓之裤襟,在这种情况下,还需要裤门襟的衣片(见图 2.8)。某些款式的女裤或宽松裤在桅缝处开门襟,常用拉链(偶尔用纽扣)闭合。每侧至少应有一片面襟和一片里襟。裤襟衣片数量取决于裤子的款式,特别取决于裤门襟是用纽扣还是用拉链扣合。

有许多裤子的款式变化不需额外的衣片但却影响织物利用率。这些款式变化包括普通的前片和打裥的前片以及直筒裤脚、小裤脚和喇叭裤脚。其他款式却需要额外的衣片,比如,某些可变换的休闲裤裤脚可以有两个或三个相应于短裤、中裤或长裤的用拉链脱卸的部分,在这样的情况下,前片和后片不得不分割。许多裤子有前结构袋,如果不是骑缝袋,可能需要从前片派生出来的袋垫(见图2.7)。裤脚翻边不需额外的衣片,但是百慕大式短裤,可能需要裤脚口。

1.2.2 裙

一般来说,一条裙子所需的衣片要比裤子所需的简单得多。最简单的裙子是无腰头的半圆裙片的圆裙,它只需一片裙片。如今许多裙子由一片裙前片和两片裙后片(左后片和右后片)组成。在这种情况下,一定有背中缝和侧缝。对于围裹裙,应该有两片前片,右前片围裹在左前片上。除了用裥或衩为裙子提供较大的张口或下摆围长,某些设计师使用三角形布片。和裤子一样,腰头和裙袋衣片可能是必须的。

许多裙子按外形轮廓进行款式变化,此外,改变腰头上部位置,改变裙的长度或使用拼腰能够产生不同的款式(见图2.8)。

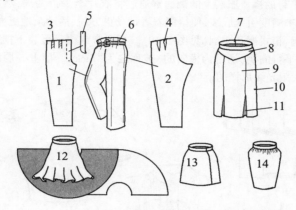

图 2.8 裤和裙的不同款式及特征示意图

1—前片 2—后片 3—裥 4—省 5—裤襻 6—蚂蟥襻 7—腰头 8—拼腰
9—前中片 10—前侧片 11—三角形布片 12—圆裙
13—围裹裙 14—上宽下窄的裙子

1.3 其他衣片和款式细节

a) 谓之"panel"的衣片。"panel"即"条形衣片",是长而窄的衣片,通常顺纵向的丝缕线裁剪。在不同服装款式中,条形衣片用于多种目的。比如,前直条用来盖住装拉链的前门襟,条形衣片还可能用来形成腰部的绳槽。称为"six-panel skirt"的六片裙由六片条形裙片构成(见图2.8,右上角裙子由六片条形裙片构成),这和称为"gored skirt"的拼片裙非常相似(见图1.8)。所谓"panel"的裙片和所谓"gore"的裙片在这里的区别是"panel"的各裙片宽度通常不一样,但"gore"的各裙片宽度是一样的。

b) 谓之"yoke"的衣片。上装上谓之"yoke"的衣片通常在颈或肩附近,有时设计用它来强调一种"强壮的肩膀"的视觉效应。衬衫(正装衬衫)上谓之"yoke"的覆肩通常包含两片缝合在一起的衣片。某些大衣也装有谓之"yoke"的过肩,看起来更"阳刚"。谓之"yoke"的衣片也可能出现在长袖的上部。

"yoke"还可以在某些裤子或裙子的腰部附近找到,谓之"拼腰"。它们的设计通常使从上臀围线到腰围线非常合体。在某些款式的裙子上,裙的宽松部分或打碎裥的部分从拼腰处下垂以产生一种丰满的下摆张开的款式效应(见图2.7)。

c) 省。典型的省是(但并非总是)在织物中特殊部位处标出的 V 形部分,这样,当 V 形部分边缘缝合在一起,平面的织物被转化成三维的形状以适合人体的三维轮廓。省的形状、大小和部位由所需的合身性及人体的部位决定。因此,省可以按它们的形状命名,如楔形省、鱼形省、弧形省、带碎裥的省和打褶的省(见图2.9)。

另外,省还可以按它们所在部位命名,如颈部省、腰省、腋下省、肩省和袖窿省等。在纸样上,省通常用锥孔和刀眼标出。前者用来标省尖,后者用来标省端。

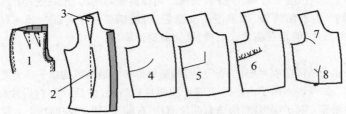

图2.9 省的示意图

1—楔形省(腰省) 2—鱼形省(腰省) 3—肩省(楔形省) 4—弧形省(腋下省)

5—折角省 6—带碎裥的省 7—袖窿省 8—打褶子的省

d) 普通的裥、碎裥和平行的裥。服装上的裥通常通过将衣片至少折两次,然后将它的一端缝入缝子而形成。裥的作用是将丰满度引入服装,使得穿着者运动更方便、活动余地更大。

裥有三种类型,即刀裥、箱式裥和阴裥(见图2.10)。裥通常或位于服装下摆或位于诸如裙子和裤子前片的腰部。多条短且平行的裥(通常为横向的)被称之为"tuck"。裙上一周沿裙长平行的裥被称为"sunray pleats",即"像太阳光线那样的裥"。

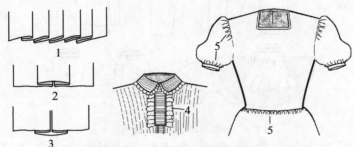

图2.10 某些裥和碎裥等的示意图

1—刀裥 2—箱式裥 3—阴裥 4—褶边 5—碎裥

碎裥和普通裥的不同之处在于碎裥中织物的折叠通常很小且无规则,而普通裥的折叠是有规则的。碎裥能够用来产生丰满度(比如在泡泡袖中)。褶边是含有碎裥的额外长度的织物,将它们缝上服装以产生波浪式的外观,以此增强女装的女性化元素。

e) 开衩。谓之"slit"或"vent"的开衩只是在衣片上的剪口。它们用于各种目的。服装下摆处的摆缝或背部的开衩可以获得较大的下摆围长,而袖口边或脚口边的开衩能够使得开口扩张,让手或脚更容易穿过。当然,某些设计师仅为款式变化使用开衩。

在衣片边缘之外的部位开衩通常称为"vent",特别是打算让空气穿过开衩处以提高穿着者生理上的舒适感时。与此不同,男式短上装常常在背部开单衩或双衩以获得款式特征。

开骑缝衩通常不需额外的毛边处理,否则,有必要对衣片进行调整,或使用额外的开衩贴边、开衩滚条等(见图2.11)。

图2.11 某些款式的开衩

1—骑缝衩 2—加贴边的衩 3—加镶片的衩 4—加滚条的衩

f)搭襻和腰带。在某些服装上使用搭襻常常是为了"支撑"穿着在身的服装。比如,某些宽松的女连衣裙被设计成带有肩襻,没有肩襻,连衣裙可能会滑下。当然,某些搭襻可能仅有装饰作用。这样,搭襻倾向于按它们所在的部位命名,比如,肩襻、袖襻和背襻。每个搭襻可能需要作为面子的衣片和作为底布的衣片,通常在它们之间加上加固用的衬头。

腰带可归为辅料范畴,但如果腰带由面料制成,必须有顺丝缕线纵向裁剪的有关衣片样板。如果腰带由蚂蟥襻握持(见图2.7和图2.8),通常没必要为蚂蟥襻准备专门的衣片,因为主衣片裁剪后剩余的碎料可足以制作蚂蟥襻。当然,必须小心,以确保在使用碎料时,它们和服装中其他的织物衣片间不存在严重的色差。

2　样板

样板是产生所需服装衣片形状的模板。使用一件服装的所有样板来绘制排料图,用以从织物上裁下衣片。对于有夹里的服装,面料和里料都需要样板,如果服装的某些部位(如领子和袋盖)需要衬头,还须有相应的样板。

对于样品服装,相同尺码且相同形状的对称衣片仅需一块样板,这时,对称的衣片可以通过水平翻转过的另一衣片的样板来描绘。因此,一件带背中缝和摆缝的上装将有左右后片和右后片,但仅需要一侧后片的样板。如果一件服装既无前中缝又不开前门襟,一般仅使用半块标明"沿前中线折"的样板以确保衣片的对称。这意味着样板将放在沿长度方向折叠的织物上,这样,将对应于服装前中线的样板边缘紧挨织物的折边对齐。其结果是,当织物被裁成样板的形状并展开时,将得到关于折叠线对称的衣片。

在准备样板时,诸如款号、衣片名称、服装尺码、待裁剪的片数等关于样板的一些信息必须标明。对于对称的折叠裁剪的样板,"CBF"或"CFF"等词也应该标出。使用刀眼标明缝头、对位记号、折叠线和省的位置,用锥孔标出省尖和纽扣的位置。刀眼应该转移到衣片上,但在批量生产中,可以在缝合服装时使用模板来标省尖和纽扣位置。

如果用作面料或里料的织物是机织的或针织的,相应于机织物经纱方向或相应于针织物纵行方向的丝缕线必须在样板上标出,这样在做排料图时,样板的丝缕线能够和织物长度方向对齐。同样,对于机织或针织的衬头,它们的样板上也需要标明丝缕线。

在未使用计算机辅助设计系统的传统的服装厂里,最初用纸做成的样板将重新用卡纸裁剪以获得更耐用的样板,并储存在样板间。对每一款式,这样的样板还可以手工推档成不同的尺码。画排料图时,拿出该款式的相应的整套卡纸样板,将它们的轮廓描绘在纸的排料图上。对于有计算机辅助设计系统的工厂,通过用数码转换板绕着样板将样板形状的信息输入计算机并储存在计算机的储存系统中。然后,使用计算机软件对它们推档以获得所需尺码范围内的所有样板。很明显,如果一款服装一个尺码有十块样板,并且该款服装订单包含十个尺码,总的样板数将达一百。使用计算机辅助设计系统,以数字形式储存样板信息将方便得多,并可以省去大量占地空间。

3　测量部位

需要制作的样板大小显然与所设计的服装要适合的人体相对应的人体测量学测量尺寸有关。关于样板设计和制作的出版物很多,因此详细描述此事不是本书的目的。第五章对此工序做了简要的介绍。不过,在本阶段读者了解这些测量尺寸(特别是在服装生产中及生产后,核查服装关键的测量尺寸)的重要性非常必要。

服装生产和检验时,许多尺寸需要测量,以确保服装和预期的尺码范围相符并确保这些尺寸在允许的容差范围内。

对于成衣上装,可能需要核查数个围长。它们是胸围(女装检验"bust girth",男装检验"chest girth")、腰围、下摆围长、袖窿、上臂围(在美国,有时被称为"muscle")和袖口大小。还可能测量数个长度或宽度,如衣长、袖长、肩宽、胸宽、背宽、胸围处宽度(不是围长)、领背宽、领深、袖口宽(如有袖口)、下摆高和背长等。如果有风帽,帽宽、从高肩点测量的帽高,帽开口长和帽顶长等可能需要加以控制。

对于裤子,可能需测量腰围、上臀围、臀围、横档、膝围和脚口大,裤长、裤脚长、前直档和后直档的长度通常需要测量。裙子的测量要简单得多,典型的测量部位有腰围、上臀围、臀围和裙长(通常为背中线长)。

对于每款成衣的检验,称为"size chart"的尺寸表(有时被称为"size specifications"或"measurement chart")按逐个尺码给出完整的参照信息。如果没有推档表,推档变量也能够从尺寸表列出的数据中推算出来。

必须注意的是,可以操作的尺寸表应该清楚并精确地规定测量部位,否则,测量会产生误差。所需进行的测量在公司和公司之间或款式和款式之间都有可能不同。

比如,袖长可能从肩点量至袖口边,也可能从后颈点(背中线和领围线交接点)量至袖口边。后者通常用来测量插肩袖。衣长可能沿背中线从后颈窝量至下摆,也可能在前中线上,从前中线和领围线交接点起量至下摆,或还可能从肩颈点("higher shoulder point",即,较高的肩点,缩写为"HSP",有时被称为"high point shoulder,即 HPS"——注意,所谓的"HSP"不一定是在肩缝上)量至前下摆。袖窿可能沿袖窿曲线形状的缝子测量,但也可能斜向从袖窿处的肩缝一端量至腋下处的袖窿底部,对于有袖服装则缝对缝,对于背心则边对边测量。曲线形状的袖窿会使精确测量胸宽和背宽显得困难。

某些公司通过明确标明测量部位的参照图来说明需要测量的尺寸,与此相反,某些公司只是用文字来对此描述,如"胸围,将在背中线和领围线交接点向下 16.5 英寸处测量"。图 2.12 简要显示了某些常用的测量部位,表 2.1 给出了图 2.12 所标编号相应的部位名称。

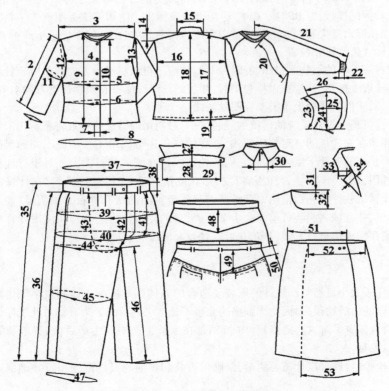

图 2.12　常用测量部位示例

表 2.1　图 2.12 所示的测量部位名称

图 2.12 中编号	测量部位名称	图 2.12 中编号	测量部位名称
1	袖口大	2	袖长
3	肩宽	4	胸宽
5	胸围	6	腰围
7	叠门宽	8	下摆围长
9	高肩点向下的衣长	10	前中线衣长
11	上臂围	12	弯量袖窿长
13	直量袖窿长(挂肩)	14	肩斜
15	颈背宽	16	背宽
17	背长	18	背中线衣长
19	下摆高	20	插肩袖前挂肩
21	背中线和领围线交接点起量的袖长	22	袖口宽
23	风帽开口长	24	从高肩点起量的风帽高
25	风帽宽	26	帽顶长
27	背中线处领高	28	背中线处领脚高
29	颈围	30	领尖间距
31	后颈深	32	前颈深
33	驳领缺嘴宽	34	驳头宽
35	从腰上缘起量的裤长	36	不计腰头的裤长
37	腰围	38	腰头宽
39	上臀围	40	臀围
41	裤襟长	42	前直裆
43	后直裆	44	横裆围长
45	膝围	46	裤脚长
47	脚口大	48	拼腰前深
49	拼腰后深	50	拼腰侧深
51	上叠片宽	52	下叠片宽
53	前叠门		

还必须注意的是,在服装生产中精确恒定的测量尺寸是无法做到的,因此,每个测量部位必须规定给予一定的容差。为了避免日后贸易争议,建议在生产前的贸易磋商中约定合理的容差。这对于新建

立业务关系的公司来说特别重要。

　　各个买主所规定的测量部位数量不一样,乍一看其理由似乎使人迷茫。在实务中,如果所有尺码的样板都由买方提供,如果相应的对等样品已经被确认,只须检验某些关键部位(如胸围、腰围、下摆围长、袖长和沿背中线衣长)应该可以了。然而,如果所有样板都由供货方制作并推档,买方可能会觉得需要检验更多的部位,特别是在买方不提供推档表而样板需要供货方推档时。在这种情况下,买方应该在尺寸表中规定尽可能多的信息,以便供货方能够从各个尺码所需的测量尺寸中推算出推档变量。

第三章　服装的织物、辅料、标签和吊牌

1　服装的织物

　　服装可以用根据样板从织物裁得的衣片,或用在平型纬编针织机上通过收针和放针技术用纱线编织的成形衣片制成。现代工艺已经有可能在全成形 V 型针床针织机和某些纬编针织圆机上直接编织无缝服装。

　　裁剪成形的服装可以有夹里或无夹里。有夹里的服装需要面料和里料,它们可以是针织的或机织的。也可以用非织造织物作为一次性使用的手术袍、防护服,或旅游者或失禁病人的内衣面料。面料可以用衬头加固以增强服装的性能、悬垂性和改善服装外观。由于衬头通常看不见的,对它们的选择强调的是成本和性能。在这一方面,倾向于最好使用非织造织物而不是机织物及针织物作为衬头。

1.1　面料

　　虽然有些服装由皮革或毛皮制成,大多数服装由纺织材料制成。许多类型的织物可以用作面料,从诸如每平方米 40 克左右的真丝电力纺的薄型织物到诸如每平方米约 700 克的粗纺毛麦尔登呢的厚型织物,它们由天然纤维、化学纤维或两者的混纺纤维制成。服装最终的价格很大程度上取决于所使用的材料以及面料的质量。面料颜色、纹理以及面料上针织的、机织的或印制的花型将成为服装吸引顾客的主要因素。无疑,面料豪华、颜色好看且引人注目的、款式高雅的服装将会被认作是高档的服装。

　　选择合适的面料时,需要考虑许多因素。比如,是用作内衣还是外衣?服装在什么气候下穿?服装是否在特殊场合穿?预期的穿着者的社会地位如何?在最终选择织物前,将不得不回答所有这些问题。

　　用作面料的织物需要具有功能性和时尚性。此外,贴近皮肤穿的织物特别需要穿着的舒适性。它们应该始终柔软、不过敏并且可能的话具有吸湿性;女士用织物可以设计成能增强其女人味的。棉和丝是用作内衣的主要天然纤维,涤纶、尼龙和弹性纱线(诸如策卡和斯潘德克斯弹力纱)是最常用的合成材料。

　　对于内衣,纬编针织物能够提供良好的透气性和延伸性。纬编平针织物(单面针织物)被广泛采用,但棉毛布可以提供更好的保暖性。其他的诸如 1+1 或 2+2 罗纹的双面针织罗纹结构延伸性更好,因此特别适合做袖口和领口以确保贴身合体。针织物有一些不足之处,相对较差的保型性和较高的缩水率是它们存在的典型问题。此外,卷边在针织物成衣裁剪时可能带来麻烦。通常平纹结构的真丝、棉和尼龙机织织物,如真丝电力纺、双绉、塔夫绸以及棉粗平布或棉细平布等也可用于内衣。由涤纶或尼龙编织的特利考经编织物以及花边在女式内衣中也非常流行。

　　外衣织物可以从更广范围的材料中选择,这取决于预期的用途。全棉织物可以用作休闲服,全毛织物可以用作正式的服装。100%的合成纤维或它们的混纺(通常和天然纤维)材料可在硬挺性、耐磨性和抗皱性等方面增强织物的性能。

外衣广泛使用机织面料,诸如府绸、有光布和派力司的平纹结构趋于最流行。某些休闲上装或长裤用咔叽布做,某些制服用华达呢或哔叽做。它们都是常用的斜纹结构。色织的或印花的方格花型和条纹花型是用作外衣的机织物的经典花型图案。花卉图案是连衣裙的流行图案。针织物也可用作面料,这主要由于它们较高的生产率,特别用在需要延伸性、保暖性和贴体合身性的用途上。然而,它们相对较差的保型性以及有起球的倾向妨碍了针织物在作为面料方面超越机织物。(起球指织物在水洗、干洗或受摩擦时,纤维的缠结在织物表面产生纤维球的现象)。

1.2 里料

有两种夹里:一种是缝在服装上的;另一种是可脱卸的,这样它可以很方便地从面子上脱下,以利于面子洗涤。有些服装可能同时有这两种夹里。

服装上加上夹里的理由之一是使它更暖和。除了提供温暖,缝上去的夹里的另一种功能是让穿着者的手方便地滑入袖子。为此,广泛使用了光滑的织物,诸如涤丝纺、春亚纺、尼丝纺和光缎羽纱等。对于冬装,化学长丝里料的丝质感可能有一种不舒服的冰凉触感,因此,有时,特别是童风衣和童夹克,可能使用机织的棉粗平布或棉绒布,或者棉的或黏胶的针织平针织物。某些尼龙休闲运动服可能使用经编针织网眼布作为里子材料,以防止尼龙黏身。

由涤纶或尼龙复丝编织成的经平绒或经绒平经编针织物曾因光滑的表面和低廉的成本被用作某些工作服的里料。

可脱卸式夹里通常用纽扣或拉链装和卸。它可用纬编割绒织物做,但如加上羽绒芯料或涤纶喷胶棉等,可脱卸式夹里为穿着者提供更多的温暖。某些服装的可脱卸式夹里用毛皮或类似的材料做成,这样的话,该服装将属协调制编码下第 43 章的类别,而不是第 61 或 62 章,即使它的面料和任何缝上去的夹里都是纺织材料。

2 辅料

辅料是几乎所有服装所必须的。了解应该如何选择辅料以及辅料和生产的关系很重要。对于金属辅料或镀金属辅料,有必要明白,大多数国家已经颁布了关于使用铅或镍的专门规定,因为它们可能对穿着者健康有害。对于用于童装的辅料,它们不得含有毒性元素,以及就如买家们通常都会要求的那样,小于 90 牛顿的力不能将它们从服装上扯下。

2.1 缝纫线

除了极少数服装,大多数衣片用缝纫线缝合。服装生产中缝纫线的主要功能是缝合、拷边、缉缝和绗缝。一般来说,待使用的缝纫线应该和将形成缝子的织物相配,线的质量应该满足适当的性能要求。

缝纫线或是短纤维纺制或是长丝的,或者由两者复合。包芯类缝纫线是一些最常用的缝纫线。它们包含了用以提供适当强度的多孔长丝芯线和保护芯线不受缝纫线和缝纫机件摩擦产生的缝纫高温影响的棉纤维包覆层,否则高温可能会将长丝熔化。传统的短纤维缝纫线用棉纺制,典型的为三股线(三根棉单纱加捻在一起),如今,最常用的短纤维纺制的缝纫线是涤纶线。

尼龙或涤纶长丝缝纫线有较高的强度和耐磨性,用在有特别要求的场合。尼龙多孔膨体或变形长丝具有较高的延伸性和回弹性,因此,多用于针织服装和泳装等。它的柔软性使得它能够很好地适用于贴身服装的包缝。单丝缝纫线也有某些应用。由诺梅克斯纺制的特殊缝纫线用于诸如消防服和赛车手套装之类潜在高温的应用场合。

缝纫线的标号定义了它的细度。对于棉纤维纺制的缝纫线,其标号定义为缝纫线总英支数的三倍。这样,以两股 40 英制支数的棉缝纫线为例,其总英制支数为 20,因此它的标号将是 60(3×20)。对于合成纤维缝纫线,使用按公制支数、特数或旦尼尔数计算的标号。在说明缝纫线规格时,了解所用的缝纫线细度表示方法非常重要。

特别对于短纤纱,为了赋予缝纫线适当的强度和韧性,施加的捻度很关键。如果过分加捻,会导致缝纫线缠结,或打结或起圈。

　　"缝合"一词指的是将一片衣片"安装"在另一片上,不过衣片也可以自己和自己缝合,就像在缝制领带时那样。因此用于缝合的缝纫线应该有足够的强度,此外,在洗烫或干洗时,它不应该比待缝合的衣片缩得更多。

　　收毛边的线迹通过用包缝线迹在织物的边缘上进行缝纫而成。在这方面,该线迹可以用于机织物和针织物。该线迹的结构是这样的:除了包覆毛边以防止它们散边外,它还是高延伸性的,这使得它能够理想地适用于内在延伸性较高的针织服装的缝纫。线迹中的弯针线通常是多孔变形长丝,它能够提供强度、延伸性和(对于内衣和泳装)所需的柔软性。

　　缉缝被用来赋予缝合衣片的缝子额外的强度,并且由于它的线迹是看得见的,它也为服装提供了装饰性的元素。它的典型应用见于衬衫领。缉缝的线不应比上缉缝的衣片缩得更多,它或者是和底色相配,即配色,或者是提供明显反差的镶色。不管哪种情况,它都必须具有较好的色牢度。

　　绗缝的作用之一是将服装中的填料或芯料固定在位。如今某些设计师已经不使用简单化的垂直、水平或方格的绗缝图案,因为计算机控制的绗缝机现在能够生产出非常精致的绗缝图案。对于厚填料,应该特别考虑绗缝线的强度,线的缩水也必须控制。线的颜色应该和工艺单规定的那样,它的色牢度也应该良好。

　　其他的线用于钉扣、锁眼和绣花。仔细按设计师给出的规格选择缝纫线极其重要。

　　在国际贸易中,服装进口商可能在服装工艺单中规定待使用的缝纫线。必须仔细考虑关于线的细节要求,以确定应该使用的线的类型和线密度。一些买家可能指定某种国际缝纫线品牌,但是应该考虑这种线是否可得到。进口线的时间和金钱花费都较大。如果可能,服装供货商可以和买方商量,是否该线能够用国产品牌的替换。线的耗用量有时会在工艺单中给出,但一般仅作参考。服装生产商常常可以在报价时根据经验估计线的耗用量,这些耗用量可以在服装打样时核实。

2.2　扣合件

　　诸如前门襟和某些袋口的开启处需要扣合件。如今在市场上可以看到各种各样的扣合件,其中的一些不仅用于扣合还用作装饰目的。

2.2.1　纽扣

　　纽扣是当今使用的最简单的扣合件。如果买方提供工艺单,需使用纽扣的款号或货号、大小、数量和颜色会在工艺单上标明。然而,实际的形状和颜色不得不通过样品来说明。

　　金属纽扣或皮扣偏贵,因此常常规定的是塑料扣。镀金属的纽扣和仿皮扣提供了相似的效果,但成本低得多。正常上色的纽扣要比诸如珠光色和大理石纹理之类特殊色泽的纽扣要便宜。

　　特别形状的纽扣将需要专门的模具。如果服装合同规定的数量大,制作这些纽扣模具的成本可以忽略。然而,如果量很小,每粒纽扣的成本将很大。作为服装供货商,和对手耐心磋商以接受某种规格,这总能找到双方有利的解决方案。可以说服买方增加服装订单中的数量,或更换纽扣,否则接受较高的服装价格。

　　必须关注纽扣的规格和现有的服装生产工艺条件。某些服装厂有两孔或四孔的订扣机,但可能没有订有脚扣的机器。某些厂可能有平头锁眼机但可能没有圆头锁眼机。如果这样的工厂在服装合同中被指定为生产商,如果服装订单中规定了这样的纽扣和纽孔,供货商履行订单时将会发现有困难,除非这些操作允许外包给有那些必要设备的生产商。对于包扣,面料将被用于生产纽扣,然而对于某些类型的织物,色差不能恰当控制,因此必须考虑使用这样的织物作包扣的可行性。此外,订在服装外面的纽扣一般需要一颗备用扣,即使没有规定,应该总是请买方确认,是否须有备用扣。

　　纽扣的大小通常用号数来定义。一号等于 1/40 英寸,即 0.635 毫米。因此一颗直径为 5/8 英寸的纽扣将被定为 25 号的(40/8 × 5)。在某些国家,纽扣的大小用它直径的毫米数表达。

2.2.2　拉链

　　拉链也是服装的重要扣合件。一根拉链的基本组成部分有拉链齿和拉链头。拉链齿沿拉链带边缘排列,而拉链头由拉链头座以及谓之"pull-tab"或"gripper"的拉柄组成。某些拉柄上有锁针以防止拉链不经意中被拉开。拉动拉柄,拉链头沿拉链齿移动,使得两根拉链带闭合或分开。在拉柄上可能有坠子以增加某些美的元素。某些拉链装双拉柄拉链头,有些有两个拉链头,这样能在两个方向拉开或闭合拉链。

根据拉链齿所用材质,拉链有三种类型。金属拉链有铜齿或铝齿的,因而它们通常较贵。通常需要拉链挡布或贴边以防止拉链直接和皮肤接触。第二种类型的拉链是尼龙拉链,它是由螺旋状的尼龙单丝盘圈形成拉链齿,因此,这类拉链可以被称为"spiral zipper(螺旋状拉链)"。它们通常是三种拉链中最便宜的。最后一类拉链是具有注塑的拉链齿的塑齿拉链,所谓的维士隆拉链就属此类。

拉链还可以分类成开尾拉链和闭尾拉链。开尾拉链用于扣合前门襟或装可脱卸式夹里或风帽。当拉链头移动到底端,开尾拉链的拉链带可以完全分开。拉链头此时留在其中一根拉链带上,该带的底端有一个针盒,用来容纳另一根拉链带底端的插针。弄清楚拉链头是在左手侧的拉链带上还是右手侧的拉链带上,这特别是对用于可脱卸式风帽上的拉链来说非常重要。由于拉链底端的尾掣,闭尾拉链不能完全拉开,它们可以用于上拉链的裤襟处,或用于扣合袋口或领圈处开口。一般要求模铸的头掣和尾掣。对于成卷供应并且必须按长度剪断的拉链,头掣和尾掣是在服装生产中安上去的。在这种情况下,将不得不使用金属爪扣的头掣和尾掣,但必须关注,以确保这些头掣和尾掣没有粗糙或锋利的边缘,应该使用拉链挡布或贴边以防止拉链的头掣和尾掣直接和皮肤接触。

对于女裙和女裤,可能使用隐形拉链。它们通常是闭尾拉链类型并且拉链闭合后,只有拉链头露出。隐形拉链不用于3岁及3岁以下的儿童穿着的服装。

图 3.1　拉链的元件
1—上掣　2—拉链头　3—拉柄　4—拉链齿　5—拉链带　6—尾掣
7—针盒销　8—针盒　9—插针　10—针片

对于尼龙拉链或塑齿拉链,拉链带、拉链齿以及拉链头的颜色通常是配色的。对于金属拉链,设计师除了应该规定拉链齿和拉链头的金属色外,还应该规定拉链带的颜色。如果有坠子,它的形状和颜色应该通过样品说明。

拉链的大小由它的长度和拉链号说明。拉链号数越高,拉链齿就越粗,能够产生的扣合力越大。4号和5号拉链通常用于外衣,2号用于薄型织物的服装。成人服装前门襟、可脱卸式风帽或可脱卸式夹里上的拉链,长度在同款服装中一般不随尺码而变。不过童风衣或童夹克前门襟的拉链长度会变化,因为对于从8岁至15岁的儿童,他们的衣长会有很大的不同。当然,没必要对每个尺码规定一个不同的拉链长度。通常的做法是将尺码分成两组或三组,每一组使用相同的拉链长度。如果这类信息在工艺单中没有说明,供货方应该要求买方澄清。

2.2.3　揿纽

人们认为,称为"snap"或"press button"的揿纽在20世纪30年代才被应用于服装。这种扣合件已被证明非常受欢迎,因为它们"噼啪"(英语中称为"snapping"或"popping",这给予揿纽"popper"的绰号,即"发噼啪声的东西"。)一下就可以扣合的便利方式。和普通纽扣相比,揿纽通常较贵,因为揿纽的主要部件通常由金属制成。一般揿纽所在部位的织物用衬头加固。

根据揿纽在服装上固定的方法,它们有两种基本类型。一种通过爪扣固定,另一种用盖固定阴扣,用销固定阳扣(见图 3.2)。对于针织物,应该使用爪扣式的,并且需要使用衬头。销钉式的一般用于厚型的或紧密的机织物。

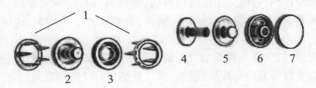

图 3.2　揿纽的组成
1—爪扣　2&5—阳扣　3&6—阴扣　4—销　7—盖

2.2.4　尼龙搭扣

英文称为"Velcro"的尼龙搭扣是一种带状扣合件的商标。它由一条带细小钩子表面的尼龙带和一条带非常柔软的圈环表面的尼龙带构成。当带钩部分压在带圈部分上,钩会被圈抓住,这样两根带子就会非常牢固地扣合在一起,并能够承受很大的力,直到它们分开。不过,如果带钩的带子从带圈的带子上剥离,两带能够非常方便地分开。尼龙搭扣像揿纽那样使用方便,但它的非金属结构使得它要便宜得多。然而必须注意,尼龙搭扣有几点不足:它的钩子倾向于聚积毛发和灰尘,它的硬钩不经意间可能会勾住结构疏松的机织物或针织物,或损坏尼龙复丝编织的拉绳。此外,有些人不喜欢拉开尼龙搭扣时发出的撕拉噪声。

2.3　衬头

衬头的主要功能之一是加固使用衬头的服装部位。和揿纽、铆钉或气眼一起使用的衬头就是为了这个目的。使用衬头的另外原因是增加衣片的硬挺性或使服装部位形状稳定。

机织或针织底布的衬头性能良好,一般也耐用。针织衬头更可能用于延伸性或弹性好的织物。由于这类衬头的丝缕线必须顺着衣片上的丝缕线,相对较低的利用率可能不能避免。另一方面,大多数非织造底布的衬头因底布中的纤维随机定向而几乎"各向同性",没有丝缕线可对。这可能大大降低裁剪损耗。在批量生产中,非织造衬头由于较低的成本和较高的利用率而广泛使用。但它们耐磨性较差的缺点阻碍了它们在需要良好耐磨性场合的应用。

虽然衬头能够缝到衣片上,但总体来说,人们使用热熔衬。它们的树脂印、涂或喷在底布上,使用平板式或传送带式黏合机,或在小规模生产时,使用蒸汽熨斗,通过施加热和压力(热融黏合)将衬头黏合到面料上。

如果买方的工艺单显示需要使用某种外国品牌的衬头,供货商仍可以和买方协商用某种国产品牌替换指定品牌的可能性。如果国产替换品具备所需的质量并且能够提供相同水平的性能,使用它们可以降低成本。不过,这种替换必须由买方事先书面确认。

2.4　填料

使用填料或芯料为服装提供额外的保暖性。谓之"filling"的芯料通常是指以无规则状态充填在服装里料和面料之间的纤维,比如羽绒芯和棉花芯。总是需要使用绗缝将服装中的这些充填纤维限制在位。谓之"wadding"的填料通常指非织造纤维层,它们可以按大小裁剪,并在服装成衣时加到面料和里料之间。作为纤维层,处理方便,这就是它在服装工业中广泛使用的原因。

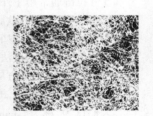

图 3.3　Lutrabond™ 纸膜

通常使用的填料有涤纶喷胶棉和腈纶喷胶棉,一般通过平方米克重来规定它们。对于非常寒冷地区穿着的冬装,比较典型的是使用每平方米 80克或以上的填料,有时袖子使用较薄的填料。某些诸如"Lutrabond™"之类极薄的非织造纸膜可用来加固填料,并且防止填料中的纤维在洗后穿透里料或面料,形成难看的纤维球。普通的每平方米 40 克的填料强度较低,最好使用纸膜加固。

2.5 其他辅料

服装中还可能使用其他的一些辅料。肩衬可以增强肩膀宽的视觉效应,给予穿着者"强壮"的形象。金属气眼和铆钉被用来产生阳刚气质,但它们现在也用于女装。服装上有时使用磁性纽扣以方便扣合。花边用来增强女人味,并常常用在女式衬衫和女式内衣上。

绳带可能会分别用在风帽、腰部或下摆上,以便绕头部或身体收紧风帽或服装。绳带的自由端或者打双折并加锁式缝固定,或者用热封、激光裁割、塑料套管(如鞋带头)固定。还可用打结固定绳带头,但不推荐在童装上使用这种方法。可以用绳头阻止绳带退入绳槽,用绳塞通过绳塞里面的弹簧帮助(弹簧型绳塞)或绳带和绳塞间的摩擦力(双眼型绳塞,见图3.4)将绳带保持在收紧状态。

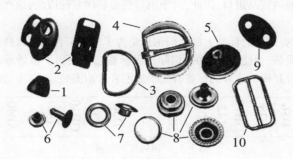

图 3.4　某些常用辅料

1—绳头　2—绳塞　3—D 型扣　4—带扣　5—有脚纽扣
6—铆钉　7—气眼　8—揿纽　9—双眼型绳塞　10—滑扣

在腰部下摆或袖口边可使用宽紧带。还可使用皮革、人造革或甚至面料制的腰带,这时腰带可能需要带扣。腰带可以用面料做的蚂蟥襻握持,否则要使用有时被称为半园扣的 D 型扣。

无论涉及什么辅料,有必要对它们明确规定。大小、形状、颜色和所用材质必须在贸易磋商中明确并予以约定。可能必须使用样品以确保有待使用的辅料的类型没有被误解。与此相似,对于颜色,配色的以及特别是指定的特殊颜色的,染色样布或具体颜色的辅料样品可能是必要的,哪怕是对于"非彩色的"辅料。

3　标签和吊牌

成衣服装上要钉标签,打吊牌。缝在服装上的标签向购买者提供了关于服装的某些信息,服装上打上吊牌是为了确认带吊牌的服装已经通过了最终服装检验。

3.1　标签

标签有许多类型,它们包括品牌标签、原产地标签、尺码标签和洗涤标签,其中大多数标签(如果不是全部)都可以在一件服装上找到。许多服装仅仅只有单个标签,给出关于品牌、产地、尺码和保养要求方面的信息。根据标签类型和服装的类型,标签位于服装不同部位。除了品牌标签,作为缝纫操作的一部分,许多标签缝在服装的缝子上,典型的例子是缝在摆缝上,这避免了单独的订标签操作。

3.1.1　品牌标签

顾名思义,品牌标签将给出服装的品牌信息。由于品牌信息可能会反映服装的"身份"以及穿着者地位,它的生产方法通常预示了品牌的地位。比如,昂贵且高级的品牌通常使用机织提花标签,并常常出现在服装的显著位置。

必须注意,注册的品牌名称和商标是拥有者的知识产权。如果供货商想用自己的品牌出口服装,他应该事先了解是否相同的品牌之前已在目的国注册。如果品牌由买方指定,供货商应该在签署合同前要求买方确认,买方是否有权这么操作。如果供货商不能获得这样的信息,他应该坚持加入关于买方产权责任的条款,比如说,"如果买方指定的品牌标签和吊牌侵犯了第三方知识产权,或造成任何争议,买方必须对后果及处理负全责"。

3.1.2 产地标签

显示服装原产国的原产地标签通常是印制的。在国际贸易中,特别是存在服装贸易限制下,给出正确的原产地信息非常重要。进口国海关税务部门从原产地信息可以确定应该按普通税率还是特别税率来征收关税。

在中国,法律要求服装出口前要经过由政府主管机构实施的标签和吊牌的强制检验以防止非法转口贸易。

如果服装和包装上没有关于原产国的信息,这将被称为"中性包装"。诸如美国、英国、德国和日本等许多国家禁止中性包装的商品进口,因此,对于出口到那些国家的服装,原产地标签是必要的。

3.1.3 尺码标签

尺码标签都是印制的。显然,标签上的尺码信息是为了使购买者能够选择合适尺寸的服装。对于目标定位于国际游客的服装,会使用所谓的"国际尺码标签",它说明了市场所在国使用的尺码以及对应的其他国家所用的相等尺码。这样的标签有助于旅游者选购服装。

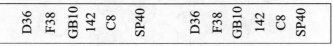

图 3.5 "国际尺码标签"

3.1.4 洗涤标签

洗涤标签或称为保养标签,它给出了如何保养服装的说明。这些说明用如图 3.6 所示意的一些国际纺织品保养标签符号的形式给出。一般使用五种符号,即洗水槽、三角、熨斗、圆圈和方块,分别表示关于洗涤、漂白、熨烫、干洗和干燥工序的建议。

保养说明通常根据面料成分确定,因此许多保养标签会显示面料的成分,通常还显示存在的任何里料成分。如果有可脱卸式夹里,它可能还需要单独的保养标签,因为对于那样的夹里可能需要专门的说明。

```
65%   Polyester  (65% 涤纶)
35%   Baumwolle/Cotton/Coton/Katoen  (35% 棉)
Hauptfutter/Lining/Doublure/Voering  (里料)
100% Polyester  (100% 涤纶)
```

图 3.6 洗涤标签

3.1.5 其他标签

除了给出品牌、原产地、尺码和保养说明信息的标签外,还可能使用提供其他信息的标签。

a)成分标签。如今越来越多的人关注面料和里料的成分。对于普通的顾客,他或她可能无法根据织物的外观和手感识别织物的成分。因此,除非这些信息已经出现在其他标签(如保养标签)上,应该要有成分标签。

b)警示标签。在某些市场,睡衣之类的服装必须具有显示服装是否满足可燃性标准并且显示"请远离火源"警示的永久性标签。对于带有小附件的童装,需要有显示噎塞窒息危险的警示标签和吊牌。这类警示可以显示在其他标签上,但有时必须使用单独的标签或吊牌。

c)生态标签。许多国家,特别是发达国家,越来越关注环境问题。它们已经制定了专门的规章并且对于通过生态问题认证的那些服装使用生态标签。有许多生态标签,最有名的一种可能是 OEKO-TEX 标准下的标签。目前,生态标签还不是强制性的,但带有生态标签的服装无疑在国际市场会非常受欢迎。

3.2 吊牌

多数吊牌用卡纸经艺术性设计制作,向顾客提供关于服装的信息。除了标签上出现的信息外,通常印有条形码,给出服装的款号、尺码和批号等信息。

如果吊牌由买方指定,关于品牌和原产地信息不应该违反他人知识产权或违反政府法规。特别需要关注指定的欧洲物品编码下的条形码(如果有的话),因为这类编码的前三位数表示的是生产国。

如果吊牌由出口商自行设计,吊牌的颜色和花型应该符合进口国的传统和习惯偏好。文字说明应该满足进口国的政府法规要求。多数进口国要求这些文字说明用它们的官方语言给出,比如对于进入加拿大市场的商品同时用英文和法文给出,对于进入中东市场的商品用阿拉伯文给出。

贸易磋商的双方应该关注对标签和吊牌的要求,即,对颜色、花型、文字说明、条形码、材质、数量以及标签和吊牌所钉的部位等要求。要求是否可行以及它们的成本都需要仔细加以考虑。

第四章 工 艺 单

1 工艺单的作用

工艺单,即"working sheet",有时又被称为"specification sheet"(缩写为 SPEC sheet)。一般来说,每个服装款式应该有一张工艺单。工艺单的作用是向服装厂的生产部门提供每件服装的详细成衣说明。工艺单上材料的描述还给予供货商关于面料、里料和辅料的规格和耗用量,以便能够估计特定服装的成本。工艺单通常由设计服装的一方提供。在服装贸易中,如果和供货商的交易按买方样品达成,买方会提供工艺单。在这种情况下,服装供货商应该仔细地阅读工艺单,因为工艺单的规定和说明成为合同的一个部分。服装厂技术人员还会根据买方工艺单的说明和打样中获取的技术参数,用自己的语言做出自己的工艺单。

工艺单的布局各公司之间总是不一样的。有的工艺单可能包含了详细的信息,而有的却非常简单。例如,如果达成的交易涉及样品,如果有关双方相互熟悉,假如供货商理解准确的要求,可能只须简要的工艺单即可。不过,如果服装款式复杂,仍然建议提供详细的信息。

2 工艺单的内容

一张工艺单应该明确说明与工艺单有关的服装款号。如果样板编号和款号不同,工艺单应该还显示样板号。这可能发生在同一套基础样板因使用(不同)的料子而产生不同的数款服装款式时。关于工艺单中必须规定的内容没有一成不变的法则,但工艺单给出的信息最好充分明确,以便能够按照要求正确地生产服装。提供一整套综合性说明的工艺单将包含以下细节,这样,关于服装应该如何制作不会有任何误解:

a)材料的规格和耗用量;

b)服装效果图;

c)技术细节;

d)尺寸表(如果不单独提供);

e)所需样板的清单(如果样板由买方提供);

f)其他必要信息,比如,标签要求和它们在服装中的位置,包装要求和包装上标志要求等。这些信息也可以另外提供。

2.1　材料的规格和耗用量

工艺单给定的材料规格和耗用量应该涉及生产该件服装的所有材料。它应该包括面料、里料、衬头、缝纫线、纽扣以及腰带之类的其他扣合件和辅料。钉扣合件和其他辅料的位置也应该给定。

a）材料规格。工艺单显示的材料类型通常规定的是统称,同时用样品形式或其他单据或协议形式提供关于材料的详细规格。比如,没有样品,光一个"ST332"的纽扣说明可能没有意义。此外,如果没有说明纱支和织物面密度等的其他单据或协议,工艺单上规定的"春亚纺"可能不足以确定有待使用的织物。

b）材料颜色。服装生产中,比较方便的是用编号而不是用名称来指出颜色。当然,所有涉及的颜色必须参照色卡或染色样布来说明。面料的颜色(服装颜色)通常在服装订单中给出,但如果涉及镶拼色,则必须通过配色表来说明颜色如何搭配,或者通过相互搭配的颜色编号的组合来为服装颜色编号。比如,如果面料主色为 2 号色,和主色 2 号色搭配的镶拼色(第二色)是 5 号色,则服装颜色编号可以为25 号色。

如果里料或辅料的颜色在工艺单中规定为"配色",即"t. i. t."（"tone in tone"的缩写）,这意味着该材料的颜色必须和该材料所在处的面料颜色相符。在这种情况下,特别当服装有镶拼色时,必须小心。如果里料或辅料被规定为特殊颜色号,这意味着,无论面料颜色如何变,该材料都使用这种特殊的颜色。

c）材料耗用量。工艺单说明的材料耗用量对应于某一特定服装尺码,通常为中档尺码。如果一张订单的尺码范围均匀分布,该订单的平均服装材料单件耗用量趋于中档尺码的耗用量。如果在贸易磋商阶段,能够提供带有关于材料耗用量正确信息的工艺单,这将有助于双方。但在该磋商阶段,并非所有工艺单都一定含有材料耗用量信息。如果这类信息不能提供,材料耗用量将不得不通过打样来确定,或凭经验来估计。

2.2　效果图

一张工艺单中可能有一张或多张被称为"working sketch"或"working drawing"的效果图。效果图描述了服装前部和背部的外观。对于上装,如果需要显示服装大身内部,肩缝不得不以拆开的形式显示,以便看得更清楚(参见本章末尾所附的工艺单中的效果图)。效果图上可包含一些简要的细节,比如,使用镶拼色的衣片、缉缝的位置及宽度等。

2.3　技术细节

技术细节为制作一件特定款式的服装提供了说明。这些细节包括:

a）每一组成部件在服装中的精确位置;

b）每一道成衣的中间工序和最终工序准确而全面的说明;

c）关于线迹类型、缝子类型、线迹密度、缝头规定,缝子毛边处理和缉缝的说明。

如果关于服装成衣的说明可以通过服装样品或其他单据确定,或可以根据相同的买家和生产商之间以往的合同中所约定的习惯做法确定,需要在工艺单中显示的技术要求可能很少。比如,某些诸如线迹类型、线迹密度、缝子类型和缝头等要求可能已经在其他单据中说明,或已经由样品确定,在这种情况下,工艺单可以不再需要提供这类信息。然而,如果既无样品,又无单据(包括工艺单)提供具体服装的技术要求,建议双方(买方和供方)最好在合同磋商中澄清并且约定,即使生产商和供货商可以仅仅遵循服装生产中使用的习惯做法操作。

2.3.1　线迹类型

虽然线迹的类型有多种,它们都是通过缝线交叉、互串或自串形成——这形成了线迹分类的基准。

在缝线交叉时,一根缝线形成的线环越过,即,绕过不同缝线形成的线环;在缝线互串时,一根缝线形成的线环从不同缝线形成的线环中穿过;在缝线自串时,一根缝线形成的线环穿过同一根缝线形成的线环(见图 4.1)。

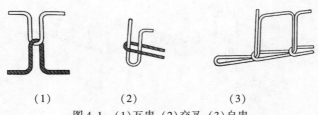

　　　　(1)　　　　　(2)　　　　　(3)
图 4.1 　(1)互串;(2)交叉;(3)自串

　　线迹被分为六个系列,每个系列含几种类型线迹。下表给出了这六个系列的特征。

表 4.1 　六个系列线迹的特征

线迹系 列号	名　称	特　点	示　例
100	链式 线迹	一线或数线形成的线环穿过织物并且通过和以 后穿过织物的线环自串固定而形成线迹	
200	(仿)手 工线迹	一线以恰当的线迹长度,从织物一面穿过至另一 面,或反之,形成线迹	
300	锁式 线迹	一线或一组线的线环穿过织物,在织物中和另一 线或另一组线交叉固定,形成线迹	
400	多线链 式线迹	一组线的线环穿过织物,通过和另一组线环交叉 及互串,形成线迹	
500	包缝 线迹	一组线的线环穿过织物,在后面的线环穿过织物 前通过自串固定;或在该组线以后再穿过织物 前,通过和一组或多组线环互串固定。至少一组 线的线环绕过织物的边缘	
600	绷缝 线迹	一组线的线环穿过已经脱落在织物表面的第二 组线的线环,然后穿过织物,和在织物底面的第 三组线的线环互串	

　　100 系列内的线迹具有极好的延伸性和光整的外观,但如果线断,很容易脱散。最常用的类型之一是 101 型,它是单线链缝,被广泛用于粗缝(即,用临时线迹缝合)。103 型(暗缝),从服装的一面是看不见的,被广泛用于缲边(见图 4.2)。

　　200 系列内的线迹在批量产生中极少被应用。为了在男式或女式度身缝制的服装上获得"手工线迹"的效果,有的缝纫机公司已经研制出手工线迹的缝纫机,它们可以用双针尖缝针生产 200 系列下的线迹。该系列中最常见的线迹类型是 209 型(见图 4.2)。

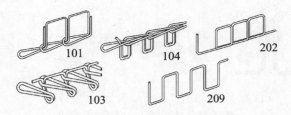

图 4.2　线迹类型示例（100 系列和 200 系列）

在 300 系列中，最广泛使用的是 301 型，通常被称为锁式缝，它不容易脱散但通常延伸性有限。它的延伸性受线迹的平衡状态影响。理想的交叉点应该发生在被缝合的织物间。不过，如果交叉点发生在织物的上面或下面，延伸性主要受制于呈平直状态的线的延伸性（见图 4.3）。线迹的平衡由相应的线上的张力确定和控制。该线迹能够在织物表面任何地方开始和完成，这使得它非常适合缝制省、裥或装贴袋等。然而，在由 301 型线迹形成的缝子的两端，通常要踩回车（前后来回缝数针）以固定缝子。广泛用于贴身内衣和泳装的 304 型和 308 型是锯齿型线迹，锯齿型的形状使得它们的延伸性比 301 型线迹大得多（见图 4.4）。

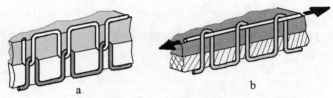

图 4.3　缝线交叉
a）正确的交叉　b）不正确的交叉

在 400 系列的线迹中，401 型，即，双线链缝（双线锁式链缝），是服装生产中第二种最广泛使用的线迹。由于较高的缝子强度和延伸性，它通常用于缝合裤子的后裆中缝，或装腰头等。404 型是锯齿型线迹，它具有和 401 型相似的线迹。406 型是多线链缝，它常被称为覆盖缝。400 系列不像 100 系列那样容易脱散（见图 4.4）。

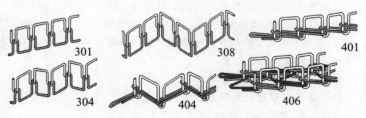

图 4.4　线迹类型示例（300 系列和 400 系列）

被称为包缝的 500 系列中的线迹广泛用于服装生产，用来包住织物修剪后的毛边以及用来缝合弹性的衣片。502 型和 503 型是双线包缝；504 型和 505 型是非常流行的三线包缝。512 型和 514 型是双针四线包缝；516 型为安全缝，它实际是 401 型线迹和 505 线迹的组合。（见图 4.5）。

600 系列的线迹是绷缝线迹，它们在缝合针织物时能提供良好的延伸性，同时具有平整及舒服的缝子外观和良好的缝子强度。某些 600 系列线迹被称为覆盖缝，但应该注意，和 400 系列的覆盖缝不同，它们同时覆盖缝子部位的面部和底部。常用的类型有 602 型、605 型和 607 型。600 系列下的多数线迹由三组线构成。一组跨接位于织物正面的平接口，第二组跨接织物反面的平接口，该两组线由第三组线穿过织物（即，缝针形成的缝针线）相互连接。602 型、605 型和 607 型分别使用两根、三根和四根缝针。(见图4.6)。

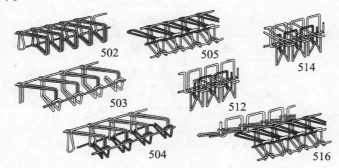

图4.5 线迹类型示例(500系列)

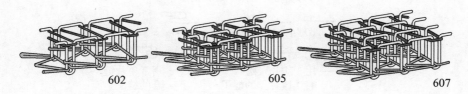

图4.6 线迹类型示例(600系列)

2.3.2 线迹密度

线迹密度是单位缝子长度上的线迹数。它根据服装成衣的具体要求确定,以便为缝子提供恰当的强度、延伸性和外观。线迹密度可以通过经验或打样决定,它取决于在成缝循环间织物向前移动的量——移动量越大,线迹密度越低。线迹密度影响了加工速度和线的耗用量,因此,影响着生产成本。线迹密度太低还可能会影响服装的质量。不同的线迹类型、不同的缝合操作以及不同的织物需要不同的线迹密度。

2.3.3 缝子类型

服装生产中缝子的主要作用是将织物部件缝合在一起。合适的缝子应该具有足够的强度,在服装穿着时所受的应力下不会被撕开;它应该具有整齐的外观,没有起皱(缝迹起皱是服装生产中常见的缝纫瑕疵之一(见第六章3.2.2节))或其他的外观上的缝纫瑕疵。此外,缝子的延伸性必须至少和织物本身相同或和相关服装部位运动量所需的相同。

缝子被分为八种类型,每一种类型中可以有许多变形。在服装生产中,常用以下四种主要缝型:

a)叠缝(第一类);

b)搭接缝(第二类);

c)包边缝(第三类);

d)对接缝(第四类)。

图4.7给出这四种主要缝型及其变形的示意。第五至第八类型的缝型也可能在服装生产中使用,图4.8给出了一些示例。请注意,这些图中的虚线表示和裁剪的边缘相反的每一织物主要部分的方向。

缝子可以用一个五位数定义。第一位数字表示缝子的类型,第二、三位定义了织物的不同形态,第四、五位表示了不同的缝针刺入及/或缝子中所用材料的影射形态。比如,最简单的叠缝会被定义为1-01-01(见图4.7中,"类型1——叠缝"标题下的第一图)。

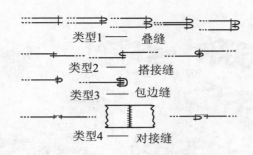

类型1 —— 叠缝

类型2 —— 搭接缝

类型3 —— 包边缝

类型4 —— 对接缝

图 4.7　第一至第四类缝型示例

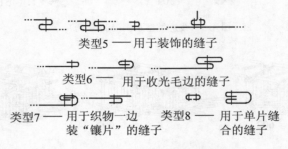

类型5 —— 用于装饰的缝子

类型6 —— 用于收光毛边的缝子

类型7 —— 用于织物一边　　类型8 —— 用于单片缝
装"镶片"的缝子　　　　　　　合的缝子

图 4.8　第五至第八类缝型示例

2.3.4　滚边和缝子毛边处理

除了衣片在编织过程中通过收放针工艺成形编织的针织服装,裁剪成形的服装在裁剪后会有毛边。毛边可能需要滚边或拷边以防止纱线从毛边滑出,否则织物会沿毛边散开,这可能破坏缝子。对于有夹里的服装,通常不必进行缝子毛边处理,因为所有衣片的毛边在面料和里料之间被包裹和保护着。对于没有夹里的服装,缝子毛边处理是必要的。

有好几种缝子毛边处理的方法,所用的方法取决于织物的类型、涉及的缝型、服装的款式和结构。拷边是广泛使用的缝子毛边处理方法,因为通过包缝线迹,衣片的毛边被很好地"捆"住,并且不会不合适地影响衣片的延伸性。滚边可以用来对服装的开口处或其边缘做毛边处理。它使用第三种缝型,用织物条(通常是斜丝缕裁剪的)沿着待滚边的织物毛边折叠包裹,将它们缝合以固定滚边。使用滚条对连衣裙领围滚边就是一个例子,使用斜丝缕的滚边包裹可脱卸式夹里所有毛边是另一个例子。用锯齿型布边剪来剪裁织物以获得锯齿型裁边被广泛用于裁剪布样。锯齿型布边裁剪还可用于不散边织物的缝子毛边处理。有些缝型,服装的毛边经双道翻折并缝合(比如6-03-01。参见图4.8中标题"类型6——用于毛边收光的缝子"下的第二图),这样,毛边被留在缝子内。这有时被称为"clean finish",即"光洁处理"。如果两片衣片的毛边折转并且嵌在缝子中,这被称为"自行光边",其中法式缝(1-06-01,图4.7中标题"类型1——叠缝"下的最后一图)就是一个典型的例子。

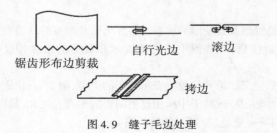

锯齿形布边剪裁　　　自行光边　　滚边

拷边

图 4.9　缝子毛边处理

2.3.5 缝头

缝头对于缝子是必要的。它被定义为在样板的边缘或样板之内所需的额外的织物量,使得织物衣片能够缝合在一起。它实际是线迹走线和织物裁边间的距离。缝头不足可能意味着在穿着受力时,线迹线会从缝子中抽出。太多的缝头会浪费织物。因此,采用的缝头影响着织物的耗用量和缝子的强度。如果样品服装的样板由买方提供,缝头会用刀眼记号标在样板上。如果样板将由服装厂制作,缝头应该在合同磋商时约定并且随后在工艺单中说明。

缝头取决于待缝合的织物的类型和结构、缝子走线的形状和缝子类型本身。和紧密的织物相比,疏松的机织或针织结构可能需要相对较大的缝头;和笔直的缝子走线相比,弧形的缝子走线通常可能要求较窄的缝头。

2.4 尺寸表

工艺单所附的尺寸表为具体的服装列出了整个尺码范围内的服装关键部位间距离的测量尺寸。它们和该服装打算适用的人群的人体测量学数据有关。尺寸表中的测量部位构成了成衣检验的准则,以便确保服装按预期尺码的正确大小来生产。

尺寸表明确规定测量要求非常重要,否则,由于测量位置的误解可能产生争议。为了对此示例,来看"背宽"的测量。横过背部或高或低的测量都可能记录到不同的长度。因此某些公司会在尺寸表中明确规定,"背宽"将在"背中线和领围线交接点向下 4 英寸处"测量。其他的公司可能提供示意图,图中明确标出测量位置。

尺寸表明确说明测量容差也非常重要。对于许多制造商,特别是服装制造商,由于材料和制作工序的实质,生产出的服装具备精确长度和宽度尺寸是不可行的。假设对于某一尺码的服装胸围规定为120cm,最后成衣胸围不大可能正好120cm。容差规定了某一具体测量尺寸的可接受范围的界限,因此它们成为接受检验的服装是应该被接受还是剔除的准则。容差的值是批量生产中所能够做得到的值和所需的合体型的准确值之间的一种折衷。

在中期检验和最终成衣检验时所需做的测量部位数目会各有不同,这特别取决于服装所需的合体性的准确度。宽松式服装要求的测量部位数目总是较度身制作的服装或紧身式服装少。另一原因可能是在所有尺码的纸样由买方提供的场合,如果买方确认了对等样品,仅仅检查几个诸如沿背中线衣长、袖长、胸围和腰围的重要尺寸,买方的技术员就能确定服装的测量尺寸是否在容差范围内,是否满足合同规定的要求。不过,如果要求服装供货商对样板推档并且买方不能提供推档表,尺寸表应该含有足够多的测量部位的测量尺寸,以便确保供货商能够推算出所需的推档变量,从而能够生产出订单规定的正确尺码和大小的服装。

3 工艺单中的缩略词和简化词

在起草工艺单时为了省时,人们常常喜欢使用容易理解的缩略词和简化词。普遍认可的缩略词通常很容易懂,但简化词有时会产生困惑,这取决于它们的派生方法。以下简化词的法则在起草工艺单时常常被遵循:

a)使用一个词组中单词的首字母,或一个复合词中组成单词的首字母,比如:

CB→Centre Back(背中线);   CF→Centre Front(前中线);

CBF→Centre Back Fold(沿背中线折叠);

CFF→Centre Front Fold(沿前中线折叠);

RS→Right Side(正面);   AH→armhole(袖窿);

BP→bust point(胸高点);   SS→shoulder slope(肩斜);

ZO→zip out(用拉链脱卸);

HSP→higher shoulder point(较高肩点)。

b)保留第一个音节或第一、第二音节,或保留第一个音节或第一、第二音节并且保留接下来音节的

第一个辅音字母,比如:

COL→colour(颜色);　　　　　　DIST→distance(距离);

Consump.→consumption(耗用量)。

c)保留一个短词的第一和最后一个字母,比如:

NK→neck(颈);　　　　　　　　LT→light(轻/淡);

DK→dark(深/暗);　　　　　　　DN→down(向下);

BK→back(背/后);　　　　　　　FM→from(从……)。

d)省略除字首的所有元音并且保留所有或必要的辅音,比如:

HV→have(有);　　　　　　　　SLV→sleeve(袖);

ACPT→accept(接受);　　　　　QNTY→quantity(数量);

PLKT→placket(门襟);　　　　　FRT→front(前/前片);

DBL→double(双道)。

e)一个或数个带相似发音的字母代替一个词,比如:

N→and(和/并且);　　　　　　　B→be(是/成为);

OZWZ→otherwise(否则)。

f)使用简化的后缀,比如:

G→-ing;　　　　　　　　　　　D→-ed;

BL→-able,-ible;　　　　　　　T,MT→-ment;

N,TN→-tion

以下的尺寸表和工艺单是服装生产中使用的尺寸表和工艺单的典型例子。公司是虚构的。

约翰及格雷斯时装公司

香港,1234 维多利亚大街 1234 号

尺 寸 表

款号:20345
商品:长裤
单位:英寸

测量部位	尺 码			
	小号	中号	大号	超大号
腰围(放松测量)	28	30	33	37
腰围(拉伸测量)	42	44	47	50
臀围(腰头下8" 处量)	42	44	47	50
腰头高	1/2	1/2	1/2	1/2
前直裆(腰头下量)	11 1/4	11 3/4	12 1/4	12 3/4
后直裆(腰头下量)	14 1/2	15	15 1/2	16
裤脚长	29 3/8	29 1/2	29 5/8	29 3/4
裤长	38 1/2	39	39 1/2	40
横裆(裆下 1" 处量)	26 1/4	27 1/2	29 1/4	31 1/2
脚口大	17 1/2	18	18 1/2	19
裤脚边高	1	1	1	1
吊环长	6	6	6	6
吊环宽	1/4	1/4	1/4	1/4

* 备注:容差和 2021 冬季订单中采用的相同。

约翰及格雷斯时装公司

香港,1234 维多利亚大街 1234 号

工 艺 单

第 1 页

季节:2021 冬季

款号:2564

商品:带用纽扣装卸的可脱卸式夹里的女式夹克衫

尺码: 8

衣长: 80cm

材料		耗用量:

面料:	门幅(cm)	耗用量(cm)
涤棉府绸,货号 2520		
Ⅰ	150	215
Ⅱ	150	80
里料:		
涤丝纺,T190		
普通部分,配色	150	120
上绗缝部分,加 40g/m².	150	60
喷胶棉,网纸膜		
绗缝间距 7cm,配色		
喷胶棉:		
40g/m² 喷胶棉	150	50
衬头:		
货号 2300	100	85

纽扣				
货号	数量	大小	颜色	部位
ST234	7	32L	t. i. t.	
加 1 颗备用扣				
ST122	17	24L	非彩色	夹里

缝线:	颜色	支数	耗用量（m）
拷边	主色	120/3	250
缝纫	主色	120/3	160
	镶色	120/3	190
锁眼	配色	80/3	90

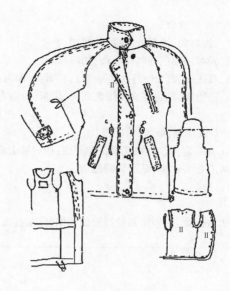

肩衬：27 号
拉链：7 号维士隆塑齿，1 根 72cm，配色
绳塞：ST56，4 颗，配色
气眼：中号，6 颗，古铜色
斜丝缕尼龙带：货号 25，3.2cm 宽，460cm 长
绣花贴片："Grace"，镶色
吊环：主色面料，10cm（成品长 6cm）

约翰及格雷斯时装公司

香港，1234 维多利亚大街 1234 号

工 艺 单

第 2 页

季节：2021 冬季

款号：2564

尺码：　　8

商品：带用纽扣装卸的可脱卸式夹里的女式夹克衫

衣长：　80cm

技术细节：

夹克：大身内部上普通涤丝纺夹里，袖子内部为上绗缝的涤丝纺夹里，带有用纽扣脱卸的可脱卸式夹里。

镶色面料：大身可脱卸式夹里；袖襻反面；袋贴边反面；门襟开纽孔直条的反面。

40g/m² 喷胶棉：前挂面；后领圈；领子；开纽孔直条；可脱卸式夹里。

上绗缝的涤丝纺：袖子内部。

内部加工：后领圈+前挂面折转；可脱卸式夹里用 0.8cm 宽的斜丝缕尼龙带滚边，用纽扣装在前挂面和后领圈处，袖窿处用扣襻固定。

面料做的绳带：2 根，0.8cm 宽，170cm 长。

纽扣间距：2.5/ 22.5/ 22.5/ 22.5cm，一颗在领襻。

可脱卸式夹里：背中线起 5/ 5/ 17/ 17/ 17/ 17cm，2 颗在袖窿处，使用扣襻。

注意：备用扣钉在左挂面下部。

叠门：

衬头：前挂面；纽孔直条（单面）；领子（双面）；袋口嵌线处；搭襻（单面）；后领圈。

前部：前中线处装明拉链。腰部绳槽用涤丝纺做，3cm 宽，从服装反面缉缝，面料做的绳带，带有绳塞，从气眼穿出。下摆处绳槽由下摆翻折，2.5cm 宽，使用面料做的绳带，在前中部固定，每侧从两个气眼穿出并且带一绳塞。

后部：腰部和下摆绳槽，参见"前部"。

下摆:下摆处夹里缝合。

袖子:两片式插肩袖,袖襻处缉缝,扣纽扣。

领子:立领,领襻用镶色面料做底面,扣纽扣。

衣袋:上袋,双嵌线袋(2 根各 0.7cm 宽),袋布:一面为主色面料,另一面为涤丝纺;

　　　下袋,贴边袋,贴边底面用镶色面料,袋布:一面为主色面料,另一面为涤丝纺。

缉缝:

净压边:绕双嵌线袋;袖窿;前门襟。

净压边加 0.7cm 宽:领子;纽孔直条和底面;插肩袖上侧缝;搭襻。

净压边加两道 0.7cm 宽:贴边袋的贴边。

2cm 宽:袖口边。

2.5cm 宽:下摆。

净压边加 2.5cm 宽:后领圈和前挂面的领围处;前挂面门襟处。

第五章　服装生产

1　生产工艺流程介绍

任何服装新款都来自于设计师的创意,但只有那些满足市场需求的款式才能够获得商业上的成功。因此,好的设计师会紧密追踪和搜集社会的、市场的以及材料方面的潮流以便预测和(如可能)影响时装的发展,并且巧妙地设计出会深受欢迎的产品。一个好的服装供货商应该拥有一支可信赖的熟知当前时装潮流和制作方法的设计团队。

要解释并且将时装效果图或服装样衣转化成商业上能够接受的服装,就需要对整个制作阶段进行仔细的工艺设计。服装制作的操作及流程因服装款式而不同,但与编织成形的服装不同,裁剪成形的服装的制作遵循着相似的工序流程。

1.1　裁剪成形服装的生产流程

裁剪成形的服装是那些由裁剪后的衣片组合而成的服装。

1.1.1　样板制作

制作流程的第一道工序是做一套样板来诠释设计师的效果图,从该套样板可以制作样衣。该工序被称为"样板制作"。虽然可以直接通过将实际用的织物或本色棉布用大头针别到胸架上来制作衣片,但是批量生产中多数都是先出一套纸样。一般,先做中等尺码样衣的纸样,由这些纸样做出的样衣被确认后,再通过被称为"推档"的工序,从该中档尺码样板制取完成服装订单所需的各个尺码的样板。

出纸样前,打板人员应该仔细地研究效果图和包含设计师规定的任何测量尺寸在内的款式特征,以弄清所需每一纸样的数量和形状。如果纸样直接从买方样衣发展而来,还必须仔细分析样衣的结构。

有经验的打板人员可能使用他们自己的公式来计算样板尺寸。对于上装,这些公式通常是基于胸围和一些特定的测量尺寸,如衣长和袖长等;对于下装,通常是基于臀围和一些特定的尺寸,如前直档和后直档等。弧形的或笔直的结构线连接纸上标出的那些点,产生各个样板。

另外一些打板人员使用原型来制作所需的样板。原型是没有任何款式特征的样板,它根据一套人体测量学数据制作,具有适合人体不同部位的具体形状。因而,有许多类型的原型,它们包括大身原型、袖片原型、裙片原型等。这些原型仅仅为打板人员提供了基本的样板形状,将这些原型样板改成最终的相应于设计师效果图或样衣并且能够提供所需的合身性的一套样板需要许多技巧和大量的经验。

正如第二章所提及的,诸如左袖和右袖之类对称的衣片或对称的左右大身,通常最初只出一块样板。最终在正常情况下,将为每一尺码制做一整套批量生产样板,每一衣片一块样板。对于一些小规模的生产,画排料图或排放那些样板以裁剪织物时,对称样板可以使用两次,即,面朝上画出一侧(比如左片)的样板,然后将样板翻转画另一侧(比如右片)样板。与此相似,如果不开前襟或没有背中缝,通常做成沿前中线(CF)或背中线(CB)对折的只有半边的前片或后片。在前中线或背中线旁将标明"CFF"或"CBF"以表示"centre front fold"即"沿前中线折",或"centre back fold"即"沿背中线折"。

样板上应该标上或剪出刀眼以标明缝头、折叠线、省的位置和对位标记等。显示裁剪前样板应该在织物上排放方向的丝缕线必须标明。最后,关于款号、样板名称、尺码等信息必须如图5.1示例的那样标出。

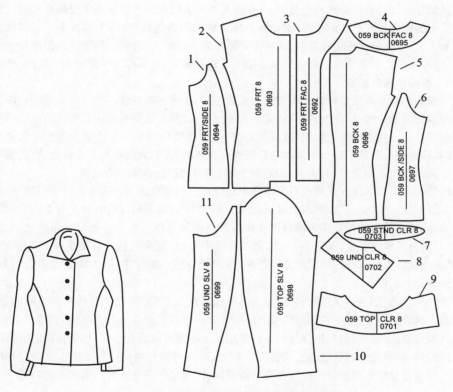

图5.1 一件服装的效果图和它的面料所需的样板

1—右前侧片 2—右前片 3—右前挂面 4—后领圈 5—右后片
6—右后侧片 7—领脚 8—底领 9—面领 10—右大袖 11—右小袖

1.1.2 样衣制作

最初的样板产生后,可以制作样衣。打样是服装制作中非常重要的一个阶段。打样除了是为了提供给买方或商家检查的样衣,批量生产所需的技术数据,诸如上黏合衬时间、上黏合衬温度、织物熨烫后的热收缩率等能够在打样中获取,有难度的成衣操作能够被识别,成衣操作所需的总时间能够从中确定。这些信息对于制定生产参数非常有用。此外,在打样中,材料和劳动力成本能够得到评估。

样衣在打样间制作,它通常装备有和缝纫车间相同的缝纫机。打样间的技术员和工人应该技术很好,经验很丰富。许多生产中的问题可能是由打样间工作人员在打样阶段发现的,这样使得大生产效率更高,成本更划算。

样衣制好后将被送出供确认。如果它们得不到确认,样板可能需要修改,样衣应该重做,直到它们被设计师或买方确认。

1.1.3 推档

在服装生产中,推档指的是将服装样板从一个尺码发展成其他尺码的技术。如果样衣被确认,将在批量生产中使用的各个尺码的样板可以由样衣的样板推档。

在推档前应该先对每一样板定义 x-y 坐标,并且推算出称为"grading rule"的推档变量。推档变量被定义为样板上每一关键点从为主的样衣样板尺寸增加或减少到订单中所需的相邻的尺码尺寸的 x 和 y 方向的增量。推档变量通常来自推档表,而推档表又是来自于基于人体测量学数据的尺寸表。

手工推档需要技巧和经验,并且劳动强度很大。因为即使一款服装,所有尺码所需的样板总数可能非常之大。

计算机辅助推档使该过程简化。先通过数码转换板绕着样板将中档尺码的样衣样板输入到计算机中。借助储存在计算机中的推档变量,每一样板的所有关键点的坐标被逐个尺码自动地计算出。通常使用数学上的样条函数产生的线条,由计算机将这些关键点连接,自动生成各个尺码的样板。如果尺码范围非常大、涉及的样板比较复杂,通常需要制作最小和最大尺码的样衣,以确保推档出来的样板之间不会走样。显然,如果为主的样衣样板直接用计算机生成,无需再将样板输入计算机。

1.1.4 织物和辅料检验

应该对织物的数量和质量检验。织物交到服装厂后,必须逐匹检验。需检验的程度取决于服装生产商对织物供货商是否能够生产出符合质量要求的织物,以及是否能通过他们自己的生产后检验来识别出疵点的信心。一些服装生产商根据以往的织物供货经验构建出一套供货商等级。如果是新的供货商,需逐匹检验的概率将较高。如果织物是折叠式包装的,在放到验布机上检验前,它们需要先绕到卷布辊上。每匹布应该编上号(打上牌),检验产生的所有数据应该做相应记录。

织物检验的主要目的之一是检查是否存在织物疵点。在发现疵点处的布边将使用牌子标出,以便在铺料时方便定位。织物的门幅和匹长应该记录下来,用于以后铺料设计时的参考。一卷织物中如果有幅宽波动,应该将最窄的幅宽作为铺料设计和画排料图的基础。需要检查每匹布的色差,如果布匹足够长,并且沿整匹织物长度,从头到尾,色调仅有微小差异,这样的色差在服装生产中能够被接受。然而,如果是短片段色差或沿织物宽度方向色差,该织物可能不能使用,因为服装上的色差不可避免,或织物利用率将非常低。

对于每一批织物,配色是必须的,特别是天然纤维的匹染织物,它们更有匹和匹之间的色差倾向。从每匹织物上剪取小样布,标上它们的匹号。所有样布一起贴近放在桌上,在自然光线下观察,然后根据它们的色调重新编组。相似门幅并且同一裁剪组别的布匹可以铺在一起,使用相同的铺料设计。

织物检验后,应该记录匹长、门幅、织物疵点数、裁剪组别编号等的信息。图 5.2 是这样的记录示例。带有严重或过量疵点数的织物应该在裁剪前退回,因为关于织物质量的索赔在裁剪后不大可能会被受理。

由于服装内在质量很大程度上取决于织物的内在质量,织物内在质量的检验也非常重要。根据织物购货合同制定的规格要求,随机取样并测试,以检查色牢度和缩水率等。如有必要,还会测定适当的织物机械性能。此外,应该做生态测试,以检查织物是否含有重金属、甲醛或农药等残余,检查它们是否满足进口国政府规定的标准。

同样,对于待使用的辅料,当它们被送到服装厂时,数量和质量的检验也是必要的。

上海东旭服装厂

上海东旭路 123 号

织物检验记录

日期:2021 年 11 月 22 日

织物货号: 2200

匹号	色号	疵点数	门幅(cm)	匹长(m)	色调组别	备注
1	4	2	110	38	1	
2	4	0	110.5	41	2	
3	4	1	110	40.5	1	
4	8	0	111	39.6	3	
5	8	0	110	38.5	3	
6	8	1	110.5	40	4	

图 5.2　织物检验记录表示例

1.1.5　铺料设计和画排料图

"lay"指的是裁剪前铺好的那堆布料。"排料图"指的是将用于织物裁剪的服装样板在单层布上的排列布局。铺料设计是制定所有各批布料如何铺放的全局计划,它包括每张排料图包含的各个尺码的样板数、排料图本身、完成订单所需的排料图数、每张排料图中每种颜色织物需要铺放的层数以及每张排料图下各个待铺放的布料长度。

在做铺料设计时,需要回答以下问题:

a) 每张排料图应该包含哪些尺码?

b) 织物应该如何铺放? 比如,每种颜色应该铺多少层? 不同颜色的织物能够一起裁剪吗?

显然,为了做出上述决定,需要来自织物检验的信息和来自买方订单的颜色/尺码/数量的搭配信息。此外,还需要考虑所涉及的织物类型、裁剪机器的工作能力以及裁剪桌的长度等。

来看图 5.3 所示的简单订单。它的每种颜色的所有尺码的服装数量是均匀分布的。画该订单排料图的一个合理的方法可能是将最小和最大尺码的样板安排在一张排料图中,次小尺码和次大尺码的样板安排在另一张中,以此类推。这样,如果 4 码和 14 码的样板包含在同一排料图中,如果在同一布层中可以铺放 40 层海军蓝色的织物和 30 层杏黄色的织物,该两尺码的所有衣片可以在一次裁剪操作中获得。

约翰及格雷斯时装公司

香港,1234 维多利亚大街 1234 号

款号:25644

订单号:H3045

日期:2021 年 11 月 22 日

颜色	尺码						总数
	4	6	8	10	12	14	
海军蓝 52	40	80	120	120	80	40	480
杏黄 46	30	60	90	90	60	30	360
总数	70	140	210	210	140	70	840

图 5.3　颜色/尺码/数量均匀分配的服装订单

当然,服装生产中的情况总是比此例要复杂得多。一张订单中的各个尺码的数量可能不是均匀搭配的,名义上同色的织物可能由于色调和幅宽的不同而不得不分成两到三组。此外,每个布层铺放的织物总层数要受织物的厚度和裁刀长度和动程的限制。膨松结构的织物铺放的层数也可能会受限制,

因为由于受上层重压,下层织物可能会变形。

最终的铺料设计只有在排料图做好后才能够确定。如果工厂有方便进行预排料的计算机辅助设计系统,如果时间许可,根据织物的门幅和预排的排料图长度最好可做几个铺料设计,然后选择最佳的一个用于生产。

定每张排料图中应该包含哪些尺码后,可以制作排料图。对于没有计算机辅助设计系统的服装厂,只能手工绘制排料图。这是个复杂的工序,通常需要技巧和经验。各个尺码的所有纸样应该先用结实耐用的卡纸重新裁制,然后应该在纸上标出相应于待裁织物的净门幅的实际排料图宽。样板逐个排放在标明的区域内,然后,用铅笔绕样板的轮廓线标出。通常先排大的样板,使它们尽可能地适合各种不同形状,尽可能紧密地排放。用划粉直接在织物上画排料图适用于单件定制的服装,但这对于批量生产来说,成本上并不划算,因此极少使用。此外,用这样的方法画排料图将导致裁剪不精确,排料图难以调整。

务必当心,要确保每块样板的丝缕线和排料图长度方向妥善对齐。对于某些织物,诸如前侧片、小袖和后侧片等的某些样板能够稍稍偏离丝缕线以改善织物利用率。但是,如果允许这样,左右样板应该给予相似并且相反方向的偏斜,以保留服装衣片最终匹配的外观。

织物的绒毛方向也不得不加以考虑。对于诸如天鹅绒之类的单向绒毛织物,所有样板应以同一方向配置。对于双向绒毛的织物,每件服装内的所有样板应该同向配置,但如果花型允许,各服装间的样板配置方向可以不同。对于素色的有光布或府绸之类的织物,不需给予方向限制。对于带有印花的、针织的或机织的图案的织物(参见图5.4),需要考虑同样的问题。比如,使用双向图案面料时,如果服装订单允许,样板可以使用双向绒毛的原则配置。随之而来的问题是,样板配置有限制的,织物利用率将较低。

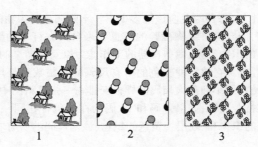

图5.4　图案的方向性
1—单向图案　2—双向图案　3—"无方向"图案

另外,对于带有印花、针织、或机织图案(特别是格子或条纹)的织物,可能有必要对花,以保证条纹或格子的线条在横过(特别是)前片时对齐。在贴袋、摆缝、背中缝以及前门襟处对花不好将产生不良的视觉效果。

手工画排料图时,对于正在绘制的整个排料图很难清楚地观察到全局,对画好的样板重新配置也不方便,因而需要有经验的工人,他们知道如何配置样板可以获得最大织物利用率。否则,织物利用率会较低,从而增加生产成本。在某些没有计算机辅助设计系统的工厂,画排料图时先在按比例缩小的区域内排放事先缩小的(通常为1⁄5)的全套样板。以这种方法,可以观察和调整整个排料图。当缩小的排料图取得满意的排料长度,可以根据缩小的排料图的配置,用原来的样板画出正式的、将用于生产的、完整大小的排料图。

如今,大多数服装厂使用了计算机辅助设计系统,排料工序变得简单得多。调出储存在计算机中的特定尺码的样板,它们在监视屏的上部以图标的形式列出或展现。在屏幕中央,标出排料图宽度,图标从代表较大的样板开始,以适当的顺序被拖放到排料图中,直到所有样板都排完。除非给出允许样板转动的指令,计算机系统将自动对齐丝缕线。如果有织物绒毛方向或图案方向的限制,操作者可以输入关于这些限制的指令,计算机将根据这些指令排放服装样板。系统将显示每张排料图所用的织物长度和织物利用率。后者通过样板本身的实际面积和排料图宽度和长度的乘积计算。样板被配置和重新配置,直到达到一个可接受的排料图长度。

排料图现在备妥待绘。建议只有在需要时才绘制排料图,特别在非常潮湿的天气,因为潮气可能会使绘好的排料图长度改变。在计算机化的裁剪中,排料图数据直接传送到裁剪工具,控制它移动并裁剪铺好的织物。

图 5.5 为一张用于裁剪的排料示例图,其中,只涉及一件服装。一般,如果一张排料图中包含的服装越多,织物利用率在正常情况下会越高,但这很大程度取决于排料图中包括的尺码、样板的形状和大小样板的相对比例。当然,用于裁剪的排料图长度还受裁剪桌长度制约。

图 5.5　图 5.1 所示样板的裁剪用排料图示例
(不受绒毛方向限制)

1.1.6　铺料

根据铺料设计的要求实施铺料。有关匹号的织物搬到裁剪桌上,根据所要求的排料长度和铺料方法放置。某些织物只能单向铺放,即,织物在铺料设备每次横移结束时需要裁断,在回程时不铺放织物。某些织物可以双向铺放,换言之,铺料设备每次横移结束时织物无需裁断,织物可以来回铺放,这可能导致较高的铺料生产率。

铺料时,织物张力需要均匀并且尽可能为最小,因为张力会造成衣片在裁剪后松弛收缩甚至变形。此外,织物层的一边必须始终对齐以保证以后所有衣片得以准确裁剪。如果存在织物疵点,织物可以在疵点处裁断,在这种情况下,当重新开始铺料时,织物必须搭接以确保跨越过裁断线的衣片是完整的。另外一种做法是可以将带有疵点的该层织物拿掉,以后使用,铺料从布层起始处重新开始。

对于针织服装,特别是针织内衣,织物通常以圆筒状而不是开幅形式铺放并裁剪。它们的裁剪排料图趋于更简单。

1.1.7　裁剪

织物铺好后,绘制好的排料图被铺放在布层的上面。顺着样板的轮廓线,裁下衣片。为了确保裁剪的边缘整齐并垂直于裁剪平面,经验和技巧仍然非常必要。关于如何开裁没有一成不变的法则,但许多操作者情愿先裁小片。

对于裁好的衣片必须检验,如果发现某些衣片走样或裁错,应该裁剪替换的衣片。需要小心,以确保新片和被调换的衣片色调相同。

1.1.8　分类、打捆和打号

裁剪后,同一尺码、同一件服装的一堆堆的衣片要进行分类,相同的衣片成"捆"放在一起,或将每一件服装所需的所有衣片"成套"放置一起。前者用于所谓的"bundle production system"(即,"成捆"作业法),衣片用手工成捆形式搬运去缝合。后者用于所谓的"unit production system"(即,"单元"作业法),服装的一套套衣片根据有待实施的操作,通过架空的、计算机控制的、机械传动的链条式搬运系统运送到各个工位。

为了避免可能的色差,常常有必要对衣片打号。同一层织物的衣片应该打上相同的号码,以提醒操作人员;从不同层织物裁下的衣片不该错误地缝合在一起。

1.1.9　上黏合衬

对于某些诸如领子、袋盖和前挂面等衣片,在缝纫前需要对它们上衬,予以加固。服装工业中常使用的衬头为热融衬,它可以通过上黏合衬工序,经过加热和加压,黏合到有关的衣片上。

在上黏合衬中,温度、压力和时间是最重要的因素。需要施加足够的热量使干性而可融的热塑性黏合剂成为半熔融状态,并且需要施加足够的压力以确保热融衬和衣片紧密接触。温度过低或压力过小会导致黏合度低,在穿着、干洗或洗涤过程中衬和布可能分层。温度过高或压力过大会导致黏合剂过分渗入衣片,产生"渗穿"的后果,从而带来较差的手感、较差的黏合度以及较差的穿着和洗涤性能。上衬时间应该足够,使得施加的温度和压力能够让衬头上的黏合剂取得所需要的熔融和渗透的效果。当然,这三个因素是相互关联的,恰当的温度、压力和时间应该在样衣制作时得以确定。

1.1.10　缝纫、小烫和修整

衣片经裁剪、分类及打号(如有必要)并且施加必要的衬头后,衣片可以送到缝工车间去缝合。

应该根据各款服装的成衣操作顺序组织生产线。由于服装成衣涉及许许多多步骤,斟酌不同款式的服装成衣操作不属本书讨论的范围。衣片的毛边可能需要收光,特别是无夹里的服装;衣片必须根据工艺单的要求使用恰当的线迹和缝子类型缝合。标签应该在缝合阶段钉上。有一些操作,比如,面领和上了衬的底领正面相对并且缝合,然后外翻(英文中称为"bagging out"),在领子夹缝到大身之前必须小烫。

做袋盖和腰头时也会采用相似的操作。下摆可能需要向上翻折并且使用缲边装置固定。可能会采用打套结来固定开口处的边角。可以使用模板标出纽扣和省的位置。

服装衣片缝合在一起形成一件服装后,缝好的服装需要修剪,以去除不必要的线头,然后需要检验。有缝纫瑕疵的服装会退回原工位,因此有可能必须换片,这将需要向裁剪车间发送加急的请求。

1.1.11　整理和成衣熨烫

检验和调整(如有必要)后,缝好的服装被送往成衣熨烫车间。

某些诸如牛仔裤和牛仔布的夹克衫可能需要洗水。洗水有许多种,包括砂洗、石洗和酵素洗的成衣洗水,但这些工序常常在专门的整理厂进行。某些服装可能需要印花或绘花。手绘和服就是后者的一个例子。如果服装厂没有技术工人和设备,这样的工作可能外包。

成衣熨烫在英文中被称为"final pressing",又可称为"top pressing"。成衣熨烫的目的是去除服装上的所有折皱,给予服装吸引人的外观,准备好装箱或上衣架。某些在成衣熨烫的高温和高压下易于损坏的辅料可能需要在成衣熨烫后再定。

1.1.12　服装检验

最后,做好的服装等待最终的检验。对于服装厂来说,一般检验服装的外观质量。成衣的外观质量可能包括做工、织物疵点、色差和测量尺寸。检验通常采用随机取样,或者有要求的可以逐件检验。

如果服装合同约定按照原样或对等样品的质量,那么样品将作为确定检查中的服装是否应该被剔除的标准。如果服装合同约定不是按样品成交,或者合同中有明确的规定,那么检验按合同约定的品质条款进行。

服装将被测量以核查如胸围、腰围、衣长和袖长等尺寸是否符合尺寸表上的数据。

如有规定,吊牌将根据买方的要求被"打"上服装。被剔除的服装需要仔细地再检查,以观察剔除的理由是否合理,若服装有理由被剔除,则通常会寻求并实施补救,除非服装疵病实在无法修整。

1.1.13　包装

服装需要包装或上衣架。虽然服装的销售合同可能规定包装条款,许多买方仅在服装生产开始前才提供详细的包装要求。这些要求也可能会包含在工艺单中,它们可能包括以下细节:

a) 服装应该如何折叠；

b) 折叠后的服装的包装尺寸；

c) 每个包装内的尺码/颜色/数量搭配；

d) 包装材料及其应用方法；

e) 包装的尺寸；

f) 包装上的标识。

包装后，服装已备妥待发。应该缮制装箱单以记录实际的装箱细节。

对于那些挂在衣架上用集装箱运输的服装，每件服装套一个塑料袋，挂在集装箱横杆上。使用绳子将衣架的钩子固定，以防止衣架在运输中从横杆上滑落。集装箱通常在厂区装箱，然后，运往集装箱堆场待运。

1.2　对于编织成形的服装

生产编织成形的服装的某些工序与生产裁剪成形的服装所采用的工序相似。不过，由于直接在针织机上从纱线编织成所需形状的成形衣片，不再需要样板设计、推档、铺料设计、画排料图以及裁剪等工序。

纱线送到工厂后需要检验，以核查它们的物理性能和外观性能是否符合规定。然后，确认后的纱线被送往针织车间。如果用于编织成形服装的纱是以绞纱形式送到工厂的，需要将它们卷绕到筒子上。络纱一般进行两次，即从绞纱到筒子纱，然后再从筒子纱到筒子纱。以这样方法络纱的目的是为了确保纱在卷装中张力均匀。现代的络纱机在每个络纱头上或者装有复杂的纱疵检测装置，或者装有简单的清纱器，以便检测并去除纱的疵点及缺陷。为使纱线更柔软，减少它们的摩擦以及防止静电荷干扰编织，可能有必要对纱线上油和上蜡。

编织衣片前，必须根据设计师的要求和效果图，或者根据买方的样衣和每一尺码成品服装的尺寸规格，仔细地对所有的技术细节进行工艺设计。

批量生产前，编织样衣以确定或核实需要的编织流程，并获得供确认的样衣。这会涉及数个工序。首先应该根据待编织纱的细度，即线密度，和针织结构所需的紧度选择合适的机号（沿针床每英寸织针数）。然后，根据衣片的最大宽度和规定的纵行密度，可以计算包括用于编织缝头在内的参与编织的织针数。根据衣片的长度、织针为特定结构而编织的线圈类型和横列密度等，机头游架往复移动的次数可以被推算出来。最后，根据衣片的形状，可以计划所需的收放针的量和次数。以这样的方法，能够确定所有衣片的编织细节，并且连同必要的说明，列明在工艺单中。为了使关于编织的规格说明更清楚，可以使用效果图。

大多数衣片在普通横机上编织，但某些衣片可以在全成形机上编织，它们通过程序设定，使用编织、集圈和不编织组合，编织所需要的花型和结构。编织规格说明中的纱线颜色和移圈等也可以用计算机辅助设计系统直接编程。在这样的系统上还可以生成花型。

遵循工艺单所给的要求，所有衣片可以被逐片织出。在编织的衣片被缝合在一起前，它们可能需要进行松弛，以释放编织中产生的内在应力和变形——这通常在蒸汽定型机上实施。对衣片的检验也还是必要的。

衣片或者（在精确线圈对线圈连接场合）在套口机上缝合，或者在针织毛衣缝合用的包缝机，或一般的绷缝机、包缝机、链缝机和锁式缝机上缝合。最新的横机纬编工艺能够编织整件的服装。这些工艺仅需最少的缝纫操作，因此大大降低了所需要缝合操作的工作量及其成本。

根据规定，缝合好的服装和整件编织的服装可能都需要经过一系列的整理工序，比如缩绒、防蛀整理、防缩整理或防起球整理，然后服装经水洗、脱水和干燥。最后常要进行蒸汽定型，这通常通过蒸汽压烫机实施。

最终的检验中，通常通过随机取样对成衣进行检验，以核实服装的尺寸并且检查发生的疵点、线圈密度和服装重量等。应该检查的参数一般取决于合同条款或原样。

编织成形的服装之包装应该遵循在贸易磋商过程中约定的包装说明。在大多数情况下,每件服装要套入一只塑料袋,然后根据具体的颜色/尺码/数量搭配装盒,接着再装纸板箱。

2 服装厂设备

现代的服装厂使用机器以便尽可能有效地完成每一操作所要求的任务。除了设计区,服装厂可以有三个主要的工作区(或称部门或者车间),即,裁剪车间、缝工车间以及整理车间。打样间在设计区内,它总是配有和缝工车间一样的机器,以便尽可能有效地设计出生产的每一阶段所需要的操作。许多机器是为了具体的操作,还为了降低缝纫操作的技术含量并且减少操作工技术和能力差异带来产品间的差异而特意制造的。

在许多工厂里,打板人员也在打样间工作,他们使用诸如打板的专用尺、曲线板、米尺、滚线轮、剪刀、刀眼夹钳、软尺和铅笔等打板工具。打样间里胸架也是必须的,用来试穿做好的样衣,或者用于出样板。

有些工厂有计算机辅助设计中心,除了计算机、打印机和绘图机外,中心应该装备数码转换板,通过它,纸样能够输入到计算机系统中。还可以使用扫描仪和数码相机以方便服装或样板设计。显然,这些任务只有在安装了必要的设计、推档和排料等软件的计算机上才能完成。

服装厂还应该有织物检验间,它通常在仓库旁。在检验间内,应该有一台或两台验布机,它们提供足够的照明,以便验布工在织物退绕并越过检验台面再卷绕时,能够方便地检测到织物疵点。某些验布机还测量织物的长度和幅宽,并为疵点部位打记号的装置。

2.1 裁剪车间设备

2.1.1 裁剪桌

在裁剪车间,织物铺放在裁剪桌上,然后裁剪、分类、打捆以及必要时打号。裁剪桌应该结实、光滑并且具有足够的长度和宽度以便实施成本效率较高的铺料,而且适用于一系列的织物。某些裁剪桌配有吸风装置以便向下握持织物或配有吹风装置以便将织物吹起,这使得铺料和裁剪操作效率更高。这些用于裁剪的桌子可能还能够方便地调节高度以方便裁剪。

2.1.2 铺料机

有许多类型的称为"fabric spreading machine"或常常被称为"cloth spreader"的铺料机,其中最简单的铺料机可能只是个安放在裁剪桌上携带织物的游架。典型的是,它们都包括支撑着卷绕在辊子上织物的游架,在游架沿裁剪桌移动时,织物越过同样也安装在游架上的张力辊退绕。这样的铺料机通常手工操作,但有些铺料机由马达驱动,以确保最小的张力均匀地施加到织物上。

在全自动的可编程的铺料机上,所有铺料功能都储存在微处理器的储存器中。关于布层的长度和铺料类型(比如,单向面朝上、单向面对面或双向来回等)用键盘输入。边缘探测器能够发出信号,控制游架横向移动,以保证所有织物层边缘很好地对齐。它们还会配有圆形刀片,在每次游架沿裁剪桌移动到终点,刀片可以横过游架移动,裁断织物。

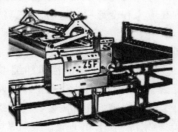

图 5.6 铺料机

2.1.3 裁剪工具

在打样间织物用剪刀裁剪,因为裁剪的通常只是单层的织物。和打样间不同,裁剪车间使用的大多数裁剪工具是电动、液压传动或气动的设备。所有裁剪设备中,直刀型电裁刀(见图5.7)是服装厂最常用的。在直刀型(电)裁刀上,用于裁剪的刀由安装在上部的马达驱动做往复运动。因为刀片通常约20毫米宽,要精确地裁剪诸如衣片上的领围线和袖窿线之类弯曲的部分常常会有难度。这样的裁刀较重,由平板下的滚子支撑,平板从待裁剪的织物下穿入。裁刀的重量和马达的振动的复合导致了裁剪的难度,使得它操作困难。不过,它的用途多样并且能够裁剪的织物类型广泛。直刀型电裁刀上有一个手柄。裁剪织物时,操作工握持手柄,沿着铺在织物上面的排料图上样板的轮廓线缓缓地推动裁刀。裁剪的织物层高可能达20多厘米,但这通常取决于具体的裁刀类型。

圆刀型电裁刀(见图5.7)是另外一种电动且手工操作的裁刀。然而,由于圆刀的直径不可能非常大,可以裁剪的层高通常约为20毫米至70毫米,并且,用它裁剪小的转角有难度。

图5.7 直刀型电裁刀(1)和圆刀型电裁刀(2)   图5.8 钢带式电裁刀

钢带式电裁刀(见5.8)专门用来裁剪衣片(特别是在小片)上弯曲的部分和小的转角,因为它能够使用细得多的金属带作为刀片。钢带由电力驱动,连续单向运动,这使得裁剪更稳定。一般,带有弯曲部分的待精确裁剪的衣片首先用直刀型电裁刀把它们分开,留着弯曲部分不修整,然后这堆"粗"割或"整垛裁下"的衣片移到钢带式电裁刀上精确裁剪。和前两种裁刀不同,在钢带式电裁刀上,操作工控制织物通过刀口而不是移动刀口穿过织物。

对于那些量足够大,需要精确形状的衣片,可以采用模具冲裁。有机械式冲裁刀和液压式冲裁刀。虽然冲裁可以得到精确裁剪的衣片,它们相对较高的成本以及制作模具的成本可能妨碍它们的应用。冲裁刀能够用来冲裁诸如领子、袋盖和贴花图案等衣片以确保形状统一。

现代工艺技术已使计算机化裁剪成为可能。裁剪头一般使用往复运动的刀口不过在一些有限的应用场合会使用激光或水压等。与绘图机上相似,织物铺放在柔韧易弯曲的尼龙刺须裁床上,在计算机控制下,裁剪头在铺好的布料中移动,逐个裁下衣片。通常在每一批铺好的布料底层先铺上打孔纸,使得布料得以很好的支撑并且使得裁床上的真空能有效发挥。在铺好的布料上面再覆上一张密封的聚乙烯膜,这样在裁剪时真空能够使布料压紧。尽管相比手工操作的裁剪工具,它的裁剪速度要快得多,但这些设备的资本成本大,必须用在大批量生产上才能够证明这样的初期费用投入的合理性。

2.1.4 上黏合衬设备

正如在本章前面部分所提及的,服装生产中广泛使用热融衬。能够施加所需温度和压力的连续式黏合机有好几种,但它们通常都有一个或两个加热区,一对轧辊和一个传送系统。将待上热融衬加固的衣片逐片面朝下地排放在黏合机的传送带上,相应的裁好的衬头放在衣片上。传送带将它们连续输送,通过加热区和轧辊(见图5.9)。衬头上的黏合剂(树脂)熔化并且将衬头黏合到衣片上。

此外,在某些服装厂可能看到非连续式的黏合设备。它们被称为平板式黏合机。其中有蒸汽压烫机,它通常用于服装的中期熨烫或者成衣熨烫(见图5.10)。衣片和裁好的衬头一起放在蒸汽压烫机的压板上受压。蒸汽提供热量,温度取决于使用的蒸汽压力。压力施加通常是机械式的或气动式的,压烫机上的定时器控制着压烫周期。

蒸汽熨斗可能是最简单的上黏合衬的工具,但是对各片衣片施加的黏合温度、压力和时间可能会不同,因此,上衬的质量难以控制并且得不到保证。

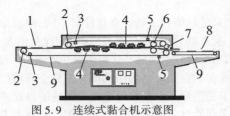

图 5.9 连续式黏合机示意图
1—上料台 2—张紧辊 3—控制辊
4—加热元件 5—清洁杆 6—轧辊
7—剥离装置 8—冷却台 9—传送带

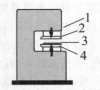

图 5.10 蒸汽压烫机示意图
1—控制面板 2—上压板
3—待黏合的衣片和衬头
4—下压板

2.1.5 裁剪车间的其他设备

裁剪车间还可能有一些其他的设备以满足特殊的技术要求。打号枪即打标签贴的枪就是其中一例,它可用来从同一层织物上裁剪下的衣片标上同一号码,该编号可以向缝纫操作工指明需要缝合在同一件服装上的衣片(即,带相同编号的衣片)。开滚条机是另外一例,它以多把用来生产直丝缕或斜丝缕滚条的圆刀为特征。这些滚条可以用来滚边或用作嵌条等。

2.2 缝工车间设备

无疑,缝工车间最重要的设备是缝纫机,它们可以按线迹的类型、涉及的缝针数、机速、用途或自动化程度分类。全自动缝纫机仅用于某些非常简单的操作,大多数缝纫操作不得不由人工完成。

2.2.1 基本缝纫机的概况

锁式缝缝纫机(平缝机)在服装厂被广泛使用。它笔直的线迹结构适合用来缝合除了具有很大弹性的织物外的几乎所有织物。如果根据所安装的缝针数分类,通常有两类锁式缝缝纫机,即,单针锁式缝缝纫机和双针锁式缝纫机。

链缝机较适合用来缝合具有弹性的织物,因为它线迹中的线环形态为缝子提供了延伸性和拉伸强度。有数种使用单针、双针或甚至多针的,生产单线、双线或多线链缝的链缝机。

包缝机主要用来修整好收光织物的毛边和缝合某些针织的或弹性的织物。包缝线迹中包边的线环形态能够将织物边缘完全包裹在内,并且同时为缝子提供良好的延伸性。包缝机主要通过参与的线数来区别。三线或四线的包缝机在服装生产中广泛使用。由五线或六线的包缝机生产的缝子实际是三线或四线包缝线迹和双线链缝线迹的结合。

绷缝机以多针多线为特征。绷缝机形成的缝子有很好的延伸性、良好的强度和整洁的外观,它们广泛用于上针织罗纹袖口或领口,以及缝合针织服装的肩缝等。

2.2.2 缝纫机的主要系统

一般来说,一台基本的缝纫机上有四个主要系统,它们是刺料系统、钩线系统、挑线系统和送料系统。

刺料系统中的主要元件是缝针,它通常是直的但在某些缝纫机上也可能是弯的。缝针经特别设计,以便在每分钟5000~8000个线迹的非常高的运转速度下能够工作良好。以一根普通锁式缝缝纫机上使用的缝针(见图5.11)为典型例子,它具有结实的针柄,以便于机器上的针夹握持。针身的一侧有一长针槽,它能够在缝针穿过织物时保护缝针线,并且确保该线不在针从织物中退出时,缝针线在针的该侧凸起。在针的另一侧,有一短针槽,它也能够在缝针下降穿过织物时保护缝针线。在针眼处有一针缺口,即"削"掉的部分,它方便了旋梭的钩尖(或,在其他线迹的场合为其他的缝纫机件)在针和线之间穿过,勾取缝针线,形成线迹。一般有两类针尖,即,球形针尖和带棱边的针尖。前者通常用于纺织材料的缝纫,后者用于皮革或类似的结实材料的缝纫。每一针尖类型又再分成许多小类,根据缝纫的织物类型来采用,以期最大程度减轻对织物的损害。

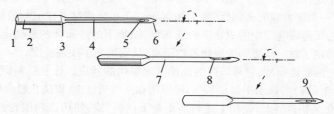

图 5.11 缝针
1—针柄头 2—针柄 3—针肩 4—长针槽 5—针眼
6—针尖 7—针身 8—针缺口 9—短针槽

钩线系统(或成缝系统)确保线环形成并且正好和钩线器交叉。在锁式缝缝纫机上,旋梭连同钩尖充当了钩线器。在诸如链缝机或包缝机之类的其他机器上,可能使用弯针和线叉(见图5.12。请注意,压脚和送布牙等没有显示)。

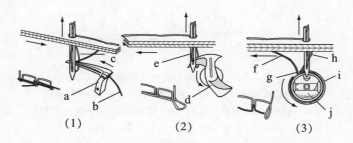

图 5.12 钩线系统示例
(1)—用于双线链缝: a)弯针 b)弯针线 c)缝针线
(2)—用于单线链缝: d)旋转钩针 e)缝针线
(3)—用于锁式缝: f)梭子线 g)梭钩尖 h)缝针线 i)旋梭 j)梭壳

在挑线系统中,广泛使用挑线杆或挑线凸轮。挑线系统的作用是在缝针下降时将缝纫线拉出线的卷装,这样能够提供足够的缝纫线形成线环;在缝针上升时它的作用是将过剩的线收回,这样能够形成平衡的线迹(见图5.13)。

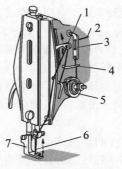

图 5.13 挑线杆、张力盘等的示例
1—挑线杆 2—缝针线 3—过线板
4—导纱钩 5—张力盘 6—缝针 7—压脚

送料系统的作用是在每一个成缝循环中向前移动织物。送料系统有几种类型,最常用的系统是下送料机构,它包含了一套锯齿状的金属牙(称为送布牙),送布牙通过支撑织物的针板上的沟槽凸起以及下降。弹簧加载的压脚将缝纫中的织物压向送布牙,织物下的送布牙的旋转运动将织物向前推动。送布牙的水平位移确定了每一线迹间的织物移动距离,从而确定了线迹长度。

差动式送料使用两套送布牙,它可以产生拉伸或起皱褶的效应。往复送料即复合送料中,缝针刺透织物时,(有时连同压脚)可以水平移动,它们可以用在某些机器上,以防止缝合中的布料层相互滑移(称为布层滑移)。在某些机器上,使用上送料而不是下送料,或,使用送料滚轮而不是送布牙来送料(见图5.14)。

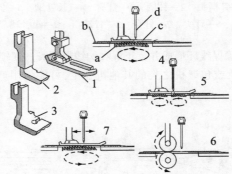

图5.14　压脚和送料机构示例
1—铰接式双趾压脚　2—单嵌线压脚　3—双嵌线压脚
4—下送料　5—差动式送料　6—送料滚轮　7—复合送料
a)送布牙　b)针板　c)压脚　d)针

除了这些系统,缝纫机上能够安装一些辅助装置以实施某些特殊功能。能够便于制作嵌线的嵌线压脚就是一个较好的例子。

2.2.3　其他设备

在缝工车间,还有一些其他类型的缝纫机。典型的有,许多服装都需要的锁眼机。它们可以分为平头锁眼机或者圆头锁眼机。和锁眼机相配的有订扣机,它们被设计成主要使用链缝线迹来订各种直径的两眼或四眼的普通圆扣。某些机器被设计成带有专门附件,以便还能够订有脚纽扣和某些揿纽。

缲边机(有时被称为暗缝缲边机)用弯针缝出链缝线迹,这样,线迹在服装的正面几乎看不出。

套结常常用于加固诸如袋口和开衩之类的服装开口的两端,而打结机就是用于此种目的的机器。打结机还可以用于钉标签。有一些设计来用于专门操作的其他机器,比如称为"jig sewing machine"的模板缝纫机(有时称为"template sewing machine"——见图5.15)以及绗缝机。前者能够准确有效地制作诸如衣袋、袖口和领子之类的衣片,后者专门设计来为带填料的服装上绗缝。某些服装厂装备有绣花机,它们是生产绣花服装或带绣花贴片的服装所必须的。

大多数现代缝纫机还装有缝针定位马达,为了缝纫具体数目的线迹并且自动停车,它能够编程。停车时,缝针可以位于预先设定的最高或最低位,压脚自动抬起。该装置可以包括传感器,以探测织物末端并且相应停车。

许多缝纫工位被设计成环绕那些半自动机器排列,以降低缝纫操作所需技能。它们和堆布装置一起工作以便自动堆放缝好的服装部件。这样的机器广泛用于那些诸如衬衫和牛仔裤之类款式几乎很少变化并且大量生产的产品。这类工位配置的例子包括自动的装衣袋的机器、领子缉缝的机器以及缝制袋盖的机器。

图 5.15　模板缝纫

图 5.16　单元作业法

2.2.4　生产线

裁缝铺一般一件服装仅由一人缝制。与它不同,在大生产中,为了生产一件服装,可能会涉及很多的缝工车间的人员。即使加工服装的一个部件,都可能需要好几种缝纫操作。很容易理解,不同的线迹可能不得不在不同的缝纫机上生产。此外,即使领子和袖子都使用锁式缝缝纫机制作,为了获得所需的产量,可能需要缝工车间的好几位缝纫工:有的准备衣领,有的准备袖子。与此相似,备好的衣领和袖子可能由另一组缝纫工装上大身,即使涉及的线迹仍然是锁式缝。缝纫工的数量取决于每单个操作所需要的时间以及缝纫工的技能。这就产生了对一个操作的标准测试值的概念。极其重要的是需要平衡好生产以确保缝工车间工作的流畅。生产的平衡是通过重视每一操作的标准测试值和缝纫工的绩效来达到的。由同一人重复相同的工作使得该人在执行该项操作中更熟练、更快,因而能提高他的绩效。由于在缝工车间,大多数操作工是计件付酬的,即,他们的薪酬和他们实施的操作数量成正比,操作工们都渴望能够提高他们的熟练程度以及操作速度。

对于大多数工厂,使用的生产方法是逐捆作业法和单元作业法,或者两者混合,而最常用的仍然是逐捆作业法。

在采用逐捆作业法的工厂里,每一个缝纫工对每一捆衣片完成一道或一系列操作后,将这捆衣片送往下一工位做进一步加工。在采用单元作业法的工厂,专门的服装搬运系统将夹有一件服装所有衣片的衣架在工位之间传送,在这些工位上,每一个操作工将完成一个操作。该操作完成后,衣架会立即被自动地发往下一工位(见图 5.16)。由于不存在需要等候一捆衣片中其余的衣片被缝制后该捆衣片才能够发出,单元作业法可有较快的产出。不过,和成捆作业法相比,它对于生产中的各工位间的不平衡承受能力较差。这样,除了对传送系统资本投入外,它需要很小心的实时生产监控以确保它的效率。正如前文提及的,混合作业法已经被研制出,依靠该作业法,某些次级成衣操作使用成捆作业法操作,最后的缝合以单元作业法操作。衣片通常由没那么复杂的手工驱动的架空轨道式传送链系统运输。

如何组织不同类型的机器以形成一条生产线是一个值得仔细考虑的问题。它一般可如下进行:首先,需要确定服装所需的线迹类型;第二,需要选定相应的机器;第三,需要确定不同操作的顺序;第四,需要计算分配给每一工位的生产任务以使生产的不平衡达到最小;最后,需要计划那些机器的布局和组织,使得从一个工位到另一工位间搬运半成品服装所花费的时间最少。

2.3　整理车间所用设备

在服装厂实施的最常见的服装整理操作是成衣熨烫。诸如成衣洗水(如砂洗、石洗或酵素洗)之类特殊的整理操作通常外包到具有所需要的专门设备和专门技术的工厂。

整理车间所用的最简单的熨烫设备是熨斗,通常为具有恒定蒸汽供应的蒸汽熨斗。带有电热元件的蒸汽熨斗能够较普通的蒸汽熨斗提供更好的熨烫效果。和蒸汽熨斗一起配套的是烫台,它需要结实并且覆有耐热衬垫,比较理想的烫台装有真空吸风装置,以有效地去除熨烫后的热量,使得服装能够安全地被搬运。用熨斗熨烫服装的不足之处在于熨烫质量完全依赖于操作工的技术。

蒸汽压烫机能够提供好得多的成衣熨烫质量,它也可以用于上黏合衬。温度、压力和熨烫时间能够预先设定,可以获得所期望的稳定而一致的质量。在服装厂的整理车间可以看到各种各样的蒸汽压烫机,以便更有效、更恒定地熨烫某些特殊的部件或达到某些特殊的目的。这些机器包括衬衫压烫机、袖口和领子的压烫机、裤子或上衣的压烫机等等,其中的一些机器像旋转木马那样(环绕)配置并且安装有自动移开压烫好部件的装置。

此外,在某些服装厂的整理部门可以看到上衣和衬衫等的人体模型整烫机(见图 5.17)、柜式整烫机(见图 5.18)、隧道式整烫机(见图 5.19)、清线头机和去污迹机等等。

图 5.17　上衣人体模型整烫机

图 5.18　柜式整烫机

图 5.19　隧道式整烫机

第六章　服装的质量问题

1　服装质量的重要性

不言而喻,质量是一件成品能被市场接受的基本先决条件。

一般来说,使用稳定的高质量的材料、上乘的管理、熟练的操作人员以及保全完好的设备等等是生产高质量服装所需的条件。然而,应该注意的是,质量问题也和生产成本紧密相关。以合理的成本生产出市场能够接受的服装是许多服装供货商所期盼的,这意味着商业上追求的质量通常指的是可接受的质量。

2　服装的质量准则

能够被认作为"可接受的"质量水平主要取决于对产品的具体要求。其主要的因素是需要符合合同规格的说明和政府的规章。第一个准则使得服装能够被买方所接受,它是在贸易磋商期间由买方和供货商共同设定的;第二个准则使得服装能够被市场接受,它是由政府(在国际贸易中,通常是进口国的政府)的立法机构所确定的法令中所设定的。对于质量的规格说明应该有一定的容差,否则,它们或者常常无法做到或者代价无法承受。如果服装质量总是趋于质量容差的上限,服装供货商将可能被认为是个好的供货商;如果质量总是趋于下限,即使质量仍然是可接受的,供货商的信誉将可能受到质疑。

在贸易磋商中,服装供货商应该仔细思考关于服装质量的合同条款,并且必须确定该质量条款按报价时所计算的成本操作是可行的。应该约定合理的容差,测试和检验的方法和样品应该在对双方所受制约的相互理解基础上约定。服装供货商应该理解,服装质量由买方工艺单、尺寸表和/或样品来定义。他应该仔细核查规格要求,如果感到不确定应该寻求澄清。比如,应该核查尺寸表中测量部位和测量容差以确保它们已经被明确定义。

服装供货商还应该确保公司的数据库关于政府对服装贸易的立法在更新。所有出口服装必须满足进口国当地强制的或规章下的标准和法规。那些规章一般和服装内在质量有关,它们是以对环境和穿着者健康和生命提供更好保护为目的而制定的。

此外,供货商应该知道,在1980年通过的联合国国际货物销售合同公约第35条(2.b)款中规定:"除当事人已另有约定的外,除非货物适合在合同成交时明示或暗示卖方的任何特定用途,除非情况表明买方并不依赖卖方的技能或判断力,或这种依赖对买方是不合理的,否则货物与合同不符"。换言之,如果买方要购买的是雨衣,即使没有详细的关于织物质量的合同条款,所提供的雨衣应该是防雨的。

3 常见的服装质量问题

服装质量可以再分为两个方面:内在质量和外在质量。熟悉一些典型的质量问题以及了解那些问题产生的原因是必要的。

3.1 内在质量

服装内在质量指的是不能够直接用眼睛观察的服装质量。它很大程度上和织物的质量有关,因此,如果织物的质量得到严格控制,对服装良好的内在质量的期盼应该是合理的。对于服装供货商,这将可能需要实施一定量的织物检验,还需要按照明示或暗示约定的标准对织物进行测试。

以下的常见问题是那些关于服装内在质量的典型问题:

a)色牢度。色牢度对顾客来说是一个敏感问题。没有人喜欢他或她的风衣长久暴露在阳光下会退色,没有人会乐意忍受漂亮的椅套沾上所坐的人的裤子颜色。色牢度有好几种,比如光照色牢度和湿摩擦色牢度等,它们按等级定义。所实施的测试取决于合同品质条款。等级数越高,色牢度越好。在很多场合,天然纤维的深色(如黑色或红色)织物有色牢度较差的倾向。染料的类型、涉及的纺织材料、染色操作以及染色参数的不同设置会影响最终的色牢度。

b)缩率。缩率是另一个敏感问题。蒸汽压烫产生的缩率可以在样衣制作时检测出,而且应该通过小心设定压烫或其他参数来防止。比如,如果在样衣制作时发现,将衬头黏合到服装前挂面上可能造成挂面衣片在长度方向收缩2%,如果上黏合衬条件保持不变,有关的纸样可能需要在长度上增加2%,以对此补偿。顾客在购买服装时,不大可能知道潜在的缩水率。虽然即使已经采用了预缩工序,可能缩率仍不可避免,但是服装大的缩率或面料和里料间不同的收缩无疑会导致顾客投诉。控制织物缩率、使用恰当的工艺以及在洗涤保养标签上提供合适的说明等都是服装供货商为将潜在缩率问题最小化应该采取的行动。

C)安全。服装的安全问题是穿着中的服装是否会影响穿着者健康或生命的问题。应该注意到,许多国家已经颁布了严格的关于服装安全性的规章,因此,了解这些规章对于服装供货商来说是很重要的。

某些安全性问题可能仅对某些人群很重要。比如,色牢度差可能会影响婴儿的健康,因为他更可能舔衣服;如果服装碰到火星,较差的阻燃性可能会很容易使体弱者处于危险中。

某些安全性问题对所有穿着者都很重要。这些问题主要由服装中某些有害的污染物的残余造成。

织物中残存的杀虫剂、甲醛、重金属,如果在可接受的极限之上,将无疑影响穿着者的健康。科学家已经发现,用22种芳香胺染色的织物可以导致皮肤癌,它们可能由偶氮染料还原分解。这就是为什么现在偶氮染料被服装进口商禁止使用的原因。此外,已经发现,某些染料,诸如颜色指数26的酸性红、颜色指数9的碱性红、颜色指数1的分散蓝等等是致癌的。某些染料,如颜色指数1的分散黄和颜色指数1的分散橙等将导致接触性皮炎。

科学研究已经揭示了以下事实:

六价铬及其化合物被归类为致癌物质,但它们极有可能被发现存在于一些厚实的织物中。镉及其化合物也是致癌的,但也可能在增塑溶胶印花、聚氯乙烯和聚氨酯织物中找到它们。五氯苯酚和2,3,5,6-四氯苯酚被用作杀虫剂或杀真菌剂,以防止真菌造成的霉变,但五氯苯酚和2,3,5,6-四氯苯酚毒性都很大并且被视作诱发癌症的物质。诸如多氯联苯和多氯三联苯之类的含氯有机染色载体主要用作杀虫剂,还用作柔软剂、染色载体和阻燃剂。已经发现,它们很容易在器官和环境中聚积,可能影响人类肝脏、荷尔蒙、免疫系统和神经系统。三丁基锡可用于抗微生物整理,但是高浓度的三丁基锡有毒,因为它会通过皮肤被吸收并且可以影响神经系统。

因此上述物质在服装生产中最好禁用或加以严格控制。

除了织物因素,服装的不良设计或成衣过程中的不良管理也可能对穿着者造成危害,留在成衣中的断针就是一个很好的例子。

还有其他的一些导致服装不安全的因素。童装上的小辅料或辅料部件可能会有窒息危险,因此,最好设计没有小部件的童装,最好确保辅料只有在合理大小的力(比如,有些买方规定为90牛顿)作用下才能够被拉下。此外,童装上不应该使用不耐用的木质的、软木的、皮革的、贝壳的或玻璃的纽扣。

d)机械性能。(服装的)重要机械性能之一是织物和缝子的强度。如果织物或缝纫线选用恰当,如果采用了恰当的技术流程,对于普通的服装几乎很少需要关注机械强度。但是,对于特殊服装必须予以特别的关注。对某些牛仔裤,可能需要测试膝部的顶破强度;对于某些工作服,可能需要测试撕裂强度。因此,机械强度是否应该控制以及需要实施哪种强度测试,将取决于合同条款和服装的预期用途。

3.2 外在质量

服装的外在质量指的是能够从服装外表评估的服装质量。它可能是顾客选购服装时考虑的主要因素。在服装厂,成衣整理工序后的服装检验的主要目的就是检查服装的外在质量。外在质量问题有许多种,它们可能源于织物的外在质量,或源于不良的做工,或源于生产技术不佳。

3.2.1 源于织物疵点的外在质量问题

服装中的织物疵点无疑会影响服装的外观。织物疵点可能因针织、机织或染整中的操作不良而产生,或可能因纤维或纱线质量不佳而产生。

诸如色花和色差等染色疵点以及油污迹、破洞和条痕等针织或机织疵点是影响服装外在质量的常见织物问题。然而应该注意,通过在服装生产的各个阶段进行仔细监控,有可能防止这些或其他的织物疵点出现在服装上。任何织物疵点在织物检验时就该检测到,如果织物疵点有可能在最终的服装上显现,检验中的织物的布边可以打上牌子,以提醒操作工在铺料操作时,采取行动去除带疵点的那段织物。除非织物发生沿匹长的短片段色差,或除非沿织物宽度方向上显现左右色差,色差不一定就是问题。通过确保同一铺料设计中包含相同色调的织物,以及通过使用同层织物裁下的衣片缝合同一件服装,能够很大程度上防止色差出现在服装上。

小的织物疵点出现在服装上是否会导致服装被剔除总是取决于它在服装上的部位。人们已经为服装不同部位设定了三个显著性等级。第一级是最显著的级别,它包括了大身前后的上部、面领、前挂面的上部、大袖、裤的前后部(除了上臀围线以上和沿内侧缝部位)。第三级是对服装外观最不显著的级别。它包括驳领的后面、前挂面下端、领子底下部分、袖子底下部分以及裤裆附近的部位。服装的其余部分属于第二显著性等级,它可能包括上装腰围线以下和下装上臀围线以上部分。小的织物疵点落在上述第二显著性等级的任何部位，服装是否能够被接受，这或多或少取决于该服装是如何穿着的。比如，在待塞入裤子的衬衫下摆附近出现的小疵点，或将被外衣上装盖住的在裤腰附近出现的小疵点，可能会被许多顾客接受。

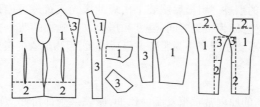

图6.1　服装外在质量的显著性等级

3.2.2　源于不良的做工或生产技术的外在质量问题

属于这一类的常见外观质量问题有：

a)源于样板的服装外在问题。一般来说，如果样衣被确认，服装中档尺码的样板就被确定了。如果整批成品服装都不符合规定的测量尺寸，有必要核查样板。采用的推档变量不对可能是主要原因。建议最好推档后对最大和最小尺码都打个样。然后按照尺寸表核查所有部位的测量尺寸，此外，应该检查服装外观以确保服装曲线部分的推档(比如袖窿和领围)是完好的。如果仅有少量服装不符合规定的尺寸，问题可能是由于缝纫操作的瑕疵，有可能是操作工没有严格遵循刀眼标出的缝头来操作。

b)源于画排料图、铺料和裁剪的服装外在问题。如果服装上格子的或条纹的花型没有正确对花，或如果服装左右侧平衡不能令人满意，排料图可能会有些问题。前者源于排板时考虑不周，后者可能是因为服装衣片的丝缕线没有对好。铺料时施加不均匀的以及过大的张力也可能导致松弛回缩，这可能会影响服装的测量尺寸。裁剪技术不佳可能会导致裁边和裁剪桌不垂直，这无疑也可能影响裁片的大小从而影响服装的测量尺寸。

c)源于缝纫的服装外在问题。常见的缝纫瑕疵有好几种，比如跳线、断线、线迹不匀、织物损坏和缝迹起皱等等。它们将影响服装的外观。

缝纫瑕疵可能源于机件配置不当。比如，线的张力设置不当或缝针的选择和安装不正确可能导致断线。压脚压力不足可能导致线迹不匀。缝纫瑕疵也可能源于机件故障。缝针或针板毛口可能造成断线。缝纫线张力过大或织物结构性的挤压可以导致缝迹起皱。此外，操作工技术不佳可以导致包括缝迹起皱在内的缝纫瑕疵。应该注意到，某些瑕疵可能源于缝料本身。缝迹起皱有时属此类，解决问题的方法之一是沿缝合线上一条衬头作为牵条，以加固缝子并增加它的抗弯刚度。

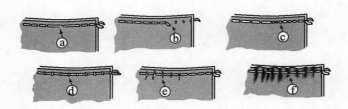

图6.2　某些典型的缝纫疵病
a)跳线　b)断线　c)线迹不匀
d)面、底线张力不平衡　e)织物损坏　f)缝迹起皱

还有一些其他的可能源于操作工的缝纫技术不佳的服装外在质量问题，它们包括诸如缉缝之类的缝子走线不均匀，或服装的左右不对称、或袖子向前或向后翘、或面领上得太紧，等等。这些问题中有一些可以通过熟练的熨烫操作补救，但其他的可能不行。因此，新工人在允许上生产线前应该经过

良好的培训。

d）源于熨斗熨烫的服装外在问题。服装通常在缝纫操作之间熨烫（英文称为"under pressing"的小烫）或缝工完成后熨烫（英文称为"top pressing"的成衣熨烫）。不良的熨烫技巧和不正确地设定熨斗的温度会在服装上留下熨烫的痕迹。熨烫后服装上出现的极光或黄斑可能是最常见的熨烫疵点。压烫不足或过分压烫也可能发生，因为熨烫时间、温度和压力以及施加的蒸汽完全取决于操作工的手烫经验。因此，如果可能，最好不要使用熨斗为服装做最后熨烫，因为成衣的最终质量将取决于熨烫操作工的技术。

必须牢记，有时外在问题的原因常常很复杂。比如，服装某个部位起皱可能是由于面料和里料不同的缩率造成的，或因为做工不好造成的等等。因而，有必要在服装生产的早期就找出确切的原因，然后设计出恰当的补救方法。就服装质量而言，牢记：防胜于治。

（说明：为了方便读者对照阅读，本书中文部分在保证英文部分原意不变的基础上，尽可能直译。读者在把握原意的基础上完全可以按照中文的修辞法去润色。）

Index to Words and Phrases

英 文	中 文	所在章节
	a	
abrasion-resistance [ə'breɪʒən-rɪ'zɪstəns]	耐摩性	第3章1.1节
absorbent [əb'sɔːbənt]	能吸湿的，能吸收的	第3章1.1节
accessory [æk'sesərɪ]	辅料	第2章1.3节
accreditation [əˌkredɪteɪʃən]	认证	第3章3.1.5节
achromatic [ˌækrəʊ'mætɪk]	非彩色的	第3章2.5节
across back	背宽	第2章3节
across chest	胸宽	第2章3节
across shoulder	肩宽	第2章3节
acrylic [ə'krɪlɪk] wadding	腈纶喷胶棉	第3章2.4节
air-blowing device	吹风装置	第5章2.1.1节
airline flight crew [kruː]	航空乘务员	第1章1.1节
A-line skirt	A 字裙	第1章1.2节
all-whether coat	防刮风下雨及日晒的大衣	第1章1.1节
anatomy [ə'nætəmɪ]	解剖学，人体，骨骼	第5章1.1.1节
anorak ['ɑːnərɑːk]	带风帽的短风衣	第1章1.1节
anthropometric [ˌænθrəʊpəʊ'metrɪk]	人体测量学的	第2章3节
anti-microbial ['æntɪ-maɪ'krəʊbɪəl]		
finishing	抗微生物整理	第6章3.1节
apparel [ə'pærəl]	服装	第3章1.1节
apparent quality	外观质量	第6章3节
applied [ə'plaɪd] collar	装上去的衣领	第2章1.1.3节
applied seam	(在织物一边)装"镶片"的缝子	第4章2.3.3节
apricot ['eɪprɪkɒt]	杏黄色	第5章1.1.5节
Arabic ['ærəbɪk] robe	阿拉伯长袍	第1章1.3节
armhole（girth）	袖窿(围长)	第2章3节
armhole ['ɑːmhəʊl]	袖窿	第1章1.1节
armhole curve	弯量袖窿	第2章3节
armhole dart	袖窿处的省	第2章1.3节
armhole straight	直量袖窿，挂肩	第2章3节
army dress blouse	(美国)军队制服上装	第1章3.2节
aromatic [ˌærəʊ'mætɪk]		
amine ['æmiːn]	芳香胺	第6章3.1节
article number	货号	第3章2.2.1节
artificial [ˌɑːtɪ'fɪʃəl] leather	人造革	第3章2.5节
assembly [ə'semblɪ]	衣片缝合	第4章2.3节
authorities [ɔː'θɒrɪtɪz]	当局	第1章3.1节
azo ['æzəʊ] dye	偶氮染料	第6章3.1节

b

back	后片	第2章1.1.1节
back neck drop	后颈深	第2章3节
back neck facing	后领圈	第2章1.1.1节
back neck width	领背宽	第2章3节
back rise	后直档	第2章3节
back sleeve	后袖片	第2章1.1.1节
back strap	背襻	第2章1.3节
back tacking ['tækɪŋ]	回车	第4章2.3.1节
back yoke depth	拼腰后深	第2章3节
bag out	(像翻口袋那样)外翻	第2章1.1.2节
balance mark	对位记号	第2章2节
band collar	立领	第1章1.1节
band knife	钢带式(电)裁刀	第5章2.1.3节
bar code	条形码	第3章3.2节
bar tack	套结,加固缝	第5章1.1.10节
bartacker ['bɑːˌtækə]	套结缝纫机,打结机	第5章2.2.3节
basic block pattern	基样,原型	第5章1.1.1节
basting ['beɪstɪŋ]	粗缝,假缝,疏缝,疏缝的针脚	第4章2.3.1节
batt [bæt]	纤维层,絮胎	第3章2.4节
batwing ['bætwɪŋ] sleeve	蝙蝠袖	第2章1.1.2节
beachwear	沙滩装	第1章1.3节
bell sleeve	钟型袖	第2章1.1.2节
bellows ['beləʊz] pocket	风箱袋	第2章1.1.4节
belt	腰带	第2章1.3节
		第3章2.5节
belt loop	蚂蟥襻	第2章1.3节
Bermuda [bə(ː)'mjuːdə] shorts	百慕大短裤	第1章1.2节
bespoke [bɪ'spəʊk] garment	定制的服装	第5章1.1.5节
bias-cut ['baɪəs-kʌt]	斜丝缕裁剪,斜裁	第4章2.3.4节
bib [bɪb]	护胸	第1章1.2节
biba ['baɪbə] dart	折角省	第2章1.3节
bikini [bɪ'kiːnɪ] bottoms	(女)比基尼短裤	第1章1.2节
to bind [baɪnd]	捆边,滚边	第4章2.3.1节
binding	滚边	第4章2.3.4节
bishop ['bɪʃəp] sleeve	主教袖	第2章1.1.2节
blazer ['bleɪzə]	西便装(式制服)	第1章1.1节
blend	混纺	第3章1.1节
blind stitch	暗缝	第4章2.3.1节
blind stitch hemming machine	暗缝缲边机	第5章2.2.3节
blouse [blaʊz]	女衬衣,(美)军上装	第1章1.1节
blouson ['bluːsɒn]	宽松式带橡筋腰的夹克	第1章1.1节
boat neck	一字领	第2章1.1.3节

bobbin ['bɒbɪn] thread	梭子线,底线	第5章2.2.2节
bobbin case	梭壳	第5章2.2.2节
bodice block	大身的基样	第5章1.1.1节
body	大身	第2章1.1节
body suit	(紧身)连短裤内衣	第1章1.3节
bottom	下装	第1章1节
bottom platen	下压板	第5章2.1.4节
bottom stop	下止,尾掣	第3章2.2.2节
bound [baʊnd] seam	包边缝	第4章2.3.3节
box pleat	箱式裥	第2章1.3节
boxer shorts	平脚裤	第1章1.2节
bra [brɑː]	文胸,胸罩	第1章1.1节
brace [breɪs]	背带	第1章1.2节
brand [brænd] label	品牌标签	第3章3.1节
brassiere ['bræsɪə]	文胸,胸罩	第1章1.1节
breeches ['brɪtʃɪz]	马裤	第1章2.1节
bridge	跨接	第4章2.3.1节
briefs [briːfs]	短内裤,三角裤	第1章1.2节
buckle ['bʌkl]	带扣,带夹	第3章2.5节
bulge [bʌldʒ] out	凸起	第5章2.2.2节
bulked continuous multifilament	膨体多孔长丝	第3章2.1节
bulky structure	膨松结构	第5章1.1.5节
bundle production system	"成捆"作业方法	第5章1.1.8节
burrs on the needle	毛针	第6章3.2.2节
bursting strength	顶破强度	第6章3.1节
bust [bʌst] (girth [gɜːθ])	女装胸围(围长)	第2章3节
bust or chest width	半胸围	第2章3节
butt	针柄头	第5章2.2.2节
butted [bʌtɪd] or flat seam	对接缝/平接缝	第4章2.3.3节
butted join	平接口	第4章2.3.1节
button ['bʌtn]	纽扣	第2章1.2.1节
button location	纽扣位置	第2章2节
buttoned-on	用纽扣脱卸的	第3章1.2节
buttonhole ['bʌtnhəʊl]	纽孔	第3章2.1节
buttonhole ['bʌtnhəʊl] machine	锁眼机	第5章2.2.3节
button-sewing machine	钉扣机	第5章2.2.3节

C

cabinet ['kæbɪnɪt] finisher	柜式整烫机	第5章2.3节
CAD system	计算机辅助设计系统	第2章2节
cadmium ['kædmɪəm]	镉	第6章3.1节
calico ['kælɪkəʊ]	白棉布,本白布	第5章1.1.1节
camisole ['kæmɪsəʊl]	(女)胸衣	第1章1.1节

cap [kæp] sleeve	帽型袖	第2章1.1.2节
capital ['kæpɪtl]		
investment [ɪn'vestmənt]	资本投入	第5章2.2.4节
carcinogen [kɑː'sɪnədʒən]	致癌物质	第6章3.1节
carcinogenic [kɑːsɪnəˈdʒɛnɪk]	致癌物(质)的	第6章3.1节
cardigan ['kɑːdɪgən]	开衫	第1章1.1节
care label	(洗涤)保养标签,(洗水唛)	第3章3.1.4节
carriage	机头游架	第5章1.2节
carrier	染色载体	第6章3.1节
carrousel [ˌkæruˈzel]	旋转(木马)式	第5章2.3节
casual ['kæʒuəl] coat	休闲大衣	第1章1.1节
casual wear [wɛə(r)]	休闲服	第1章1.1节
category ['kætɪgərɪ]	类别	第1章1.1节
category number	类别号	第1章2.2节
CB length	背中线长	第2章3节
centre back fold (CBF)	沿背中线折,无背中缝	第2章2节
centre back seam	背中缝	第2章1.1.1节
centre front (CF)	前中线	第2章2节
centre front fold (CFF)	沿前中线折,无前中缝	第2章2节
centre front neck	领圈前中线处	第2章3节
centre seat seam	裤后裆中缝	第4章2.3.1节
CF length	沿前中线衣长	第2章3节
chain stitch machine	链缝机	第5章2.2.1节
chainstitching machine	链缝机	第5章1.2节
chest (girth)	(男装)胸围(围长)	第2章3节
chest [tʃest] pocket	胸袋,(西装)手巾袋	第2章1.1.4节
children's coat	童风衣	第1章1.3节
chinos ['tʃɪnəʊz]	咔叽休闲裤	第1章1.2节
chintz [tʃɪnts]	有光布,轧光布	第3章1.1节
chlorinated organic carrier	含氯的有机染色载体	第6章3.1节
choking ['tʃəʊkɪŋ]		
hazard ['hæzəd]	窒息危险	第3章3.1.5节
chromium ['krəʊmjəm] (VI)	六价铬	第6章3.1节
CI acid Red 26	颜色指数26的酸性红	第6章3.1节
CI basic Red 9	颜色指数9的碱性红	第6章3.1节
CI Disperse Blue 1	颜色指数1的分散蓝	第6章3.1节
CI disperse Orange 1	颜色指数1的分散橙	第6章3.1节
CI disperse Yellow 1	颜色指数1的分散黄	第6章3.1节
circular ['sɜːkjʊlə] sleeve	披肩短袖	第2章1.1.2节
circular ['sɜːkjʊlə] weft knitting		
machine	纬编针织圆机	第3章1节
claim	索赔	第5章1.1.4节
clean finish	光整处理	第4章2.3.4节

cleaning bar	清洁杆	第5章2.1.4节
closed zipper	闭尾拉链	第3章2.2.2节
cloth spreaders	铺料机	第5章2.1.2节
coat	风衣,大衣	第1章1.1节
coin [kɒɪn] pocket	零钱袋	第2章1.1.4节
collar ['kɒlə]	衣领	第1章1.1节
collar CB depth	领子背中线处高	第2章3节
collar depth [depθ]	领高	第2章1.1.3节
collar notch [nɒtʃ]	驳领缺嘴宽	第2章3节
collar press	压领机	第5章2.3节
collar spread [spred]	领尖间距	第2章3节
collar stand	领脚	第1章1.1节
collar stand CB depth	领脚背中线处高	第2章3节
collarless	无领的	第1章3.2节
colour card	色卡	第4章2.1节
colour fastness ['fɑːstnɪs]	色牢度	第3章2.1节
colour fastness to light	光照色牢度	第6章3.1节
colour fastness to wet rubbing	湿摩擦色牢度	第6章3.1节
colour matching	配色	第5章1.1.4节
colour shading	色差	第2章1.3节
colour spot	色花	第6章3.2.1节
colour swatch [swɒtʃ]	色样,染色样布	第3章2.5节
colour tone	色调	第5章1.1.4节
colour/ size/ quantity assortment	颜色/ 尺码/ 数量搭配	第5章1.1.5节
colour-matching table	配色表	第4章2.1节
Committee for the Implementation of Textile Agreements (CITA)	纺织品协议执行委员会	第1章3.1节
compatible [kəm'pætəbl] size	般配的尺码	第1章2.1节
competent government authority	政府主管部门	第3章3.1.2节
composition [kɒmpə'zɪʃən]	成分	第1章2.1节
composition label	成分标签	第3章3.1.5节
compound ['kɒmpaʊnd]	化合物	第6章3.1节
compound ['kɒmpaʊnd] feeding	复合送料	第5章2.2.2节
computer memory	计算机储存	第2章2节
computer-aided grading	计算机辅助推档	第5章1.1.3节
concealed [kən'siːld] zipper	隐形拉链	第3章2.2.2节
concentration [ˌkɒnsən'treɪʃən]	浓度	第6章3.1节
cone	(圆锥型)纱筒,纱的筒子	第5章1.2节
consumption [kən'sʌmpʃən]	耗用量	第3章2.1节
contact dermatitis [ˌdɜːmə'taɪtɪs]	接触性皮炎	第6章3.1节
container	集装箱	第5章1.1.13节
container yard	集装箱堆场	第5章1.1.13节
continuous fusing machine	连续式黏合机	第5章2.1.4节

contour ['kɒntʊə]	轮廓	第2章1.3节
contractual quality terms	合同的品质条款	第5章1.1.12节
contractual specifications	合同规格说明	第6章2节
contrasting [kən'trɑːstɪŋ] colour	镶色	第3章2.1节
control panel	控制面板	第5章2.1.4节
control roller	控制辊	第5章2.1.4节
be conversant with [kən'vɜːsənt]	对……熟悉的	第5章1节
conveying system	传送系统	第5章2.1.4节
conveyor belt	传送带	第5章2.1.4节
conveyorised [kən'veɪəraɪzd] press	传送带式连续压烫机	第3章2.3节
cooling station	冷却台	第5章2.1.4节
cord [kɔːd] tunnel ['tʌnl]	绳槽	第2章1.3节
cord end	绳头	第3章2.5节
cord piping	嵌线	第5章2.2.2节
cord stop	绳塞,卡扣	第3章2.5节
core-spun [kɔː-spʌn] sewing thread	包芯纺缝纫线	第3章2.1节
corset ['kɔːsɪt]	(女)束腰塑身内衣	第1章1.1节
cotton	棉	第1章2.2节
cotton flannel ['flænl]	棉绒布	第3章1.2节
cotton sheeting ['ʃiːtɪŋ]	棉粗平布	第3章1.1节
cotton shirting ['ʃɜːtɪŋ]	棉细平布	第3章1.1节
counter sample	对等样品(通常指卖方按买方原样制作的样品)	第5章1.1.12节
country of origin ['ɒrɪdʒɪn]	原产国	第1章2.3节
course density	横列密度(纵密)	第5章1.2节
cover stitch	覆盖缝	第4章2.3.1节
crease-resistance [kriːs-rɪ'zɪstəns]	抗皱性	第3章1.1节
crepe de chine [kreiːpdə 'ʃiːn]	双绉	第3章1.1节
criterion [kraɪ'tɪərɪən] (pl. criteria [kraɪ'tɪərɪə])	准则,标准	第1章2.3节
crocheted ['krəʊʃeɪd]	钩编的	第1章2.1节
cuff ['kʌf]	袖口,袖克夫	第1章1.1节
cuff depth	袖口(克夫)宽	第2章3节
cuff press	袖口压烫机	第5章2.3节
culottes [kjʊ(ː)'lɒts]	裙裤	第1章1.2节
cup seaming machine	(针织毛衣用的)包缝机	第5章1.2节
curling ['kɜːlɪŋ]	卷边	第3章1.1节
curved dart	弧型省	第2章1.3节
customary ['kʌstəmərɪ] practice	习惯做法	第4章2.3节
Customs and Excise [ek'saɪz]		

Office	海关税务机构	第1章2节
cutaway ['kʌtəweɪ]	下摆圆角的，一部分切掉的	第1章2.1节
cut-fashioned	裁剪成形	第1章2.3节
cutting blade	裁刀	第5章1.1.5节
cutting point	棱边针尖	第5章2.2.2节
cutting table	裁剪桌	第5章1.1.5节

d

damage to fabric	织物损坏	第6章3.2.2节
dart [dɑːt]	省	第2章1.3节
dart end	省尖	第2章1.3节
dart point	省尖	第2章2节
dart sides	省端	第2章1.3节
dart site	省的位置	第5章1.1.1节
decorative ['dekərətɪv] seam	装饰缝	第4章2.3.3节
defect	疵病	第5章1.2节
de-lamination [diˌlæmɪˈneɪʃən]	分层，脱层	第5章1.1.9节
to de-skill	降低技能要求	第5章2节
denier ['denjə]	旦尼尔(旦数，纤度)	第3章2.1节
denim jackets	牛仔布夹克	第5章1.1.11节
design	花型，设计	第1章1.1节
detachable [dɪˈtætʃəbl] hood	可脱卸风帽	第2章1.1.3节
dewater [diːˈwɔːtə]	脱水	第5章1.2节
diagonal [daɪˈægənl]	斜向的，斜丝缕的	第2章1.1.3节
die cutter	(模具)冲裁刀	第5章2.1.3节
die cutting	用模具冲裁	第5章2.1.3节
differential [ˌdɪfəˈrenʃəl] feeding	差动式送料	第5章2.2.2节
digital ['dɪdʒɪtl] camera	数码相机	第5章2节
digitising ['dɪdʒɪtaɪzɪŋ]	通过数码转换板输入	第2章2节
digitizer ['dɪdʒɪtaɪzə]	数码转换板	第5章2节
dinner jacket	(没有燕尾的)晚礼服(上装)	第1章1.1节
dinner suit	正餐礼服套装	第1章1.3节
discontinuous [ˌdɪskənˈtɪnjuəs]		
fusing equipment	非连续式黏合设备	第5章2.1.4节
disposable [dɪsˈpəʊzəbl] garment	一次性服装	第3章1节
distorted	变形，走形	第5章1.1.5节
divided [dɪˈvaɪdɪd] skirt	裙裤	第1章1.2节
Dolman ['dɒlmən] sleeve	多尔曼袖	第2章1.1.2节
double breasted	双排纽	第1章1.1节
double jersey	双面针织物	第3章1.1节
double piping foot	双嵌线压脚	第5章2.2.2节
double-needle four-thread		
overlock stitch	双针四线包缝	第4章2.3.1节

double-needle lockstitch machine	双针平缝机	第5章2.2.1节
double-pointed needle	双针尖缝针	第4章2.3.1节
double-tag slider	双拉柄拉链头,双柄拉头	第3章2.2.2节
double-thread chain stitch		
(double locked chain stitch)	双线链缝	第4章2.3.1节
double-turned	双道翻折	第4章2.3.4节
down [daʊn] filling	羽绒填芯	第3章1.2节
drag	拖拽	第5章1.1.5节
drape [dreɪp]	悬垂性	第3章1节
drawstring [ˈdrɔːstrɪŋ]	(用于收紧的)绳带	第1章1.1节
		第3章2.2.4节
dress	连衣裙	第1章1.1节
dress shirt	正装衬衫	第1章1.1节
dress stand	胸架,人形模型	第5章1.1.1节
D-ring	半园扣	第3章2.5节
drop feed mechanism	下送料机构	第5章2.2.2节
dropped waist	低腰	第1章1.2节
dry clean	干洗	第3章2.1节
duster	防尘服	第1章1.3节
duster [ˈdʌstə] coat	防尘大衣,风衣	第1章1.1节
duties [ˈdjuːtɪz]	关税	第1章2.3节
dyeing parameter	染色参数	第6章3.1节
dyes	染料	第6章3.1节

e

ecological [ˌekəˈlɒdʒɪkəl] label	生态标签	第3章3.1.5节
ecological [ˌekəˈlɒdʒɪkəl] test	生态测试	第5章1.1.4节
edge detector	探边器	第5章2.1.2节
edge neatening	拷边(收毛边)	第3章2.1节
edge seaming	包缝,锁边缝	第4章2.3.1节
elastic band	宽紧带	第3章2.5节
elastic recovery	回弹性	第3章2.1节
elastomeric	弹性体的,有弹力的	第3章1.1节
elbow [ˈelbəʊ] length sleeves	中袖	第2章1.1.2节
electrically powered equipment	电动设备	第5章2.1.3节
electrostatic [ɪˈlektrəʊˈstætɪk]		
charge	静电荷	第5章1.2节
elegant [ˈelɪgənt] style	高雅的款式	第3章1.1节
embroidering [ɪmˈbrɔɪdərɪŋ]	绣花	第3章2.1节
embroidering [ɪmˈbrɔɪdərɪŋ]		
machine	绣花机	第5章2.2.3节
embroidery patch	绣花贴片	第5章2.2.3节
engineer	工艺设计	第5章1节

ensemble [ɑːnˈsɑːmbl]　　　便服套装　　　　　　　第1章1.3节

entry summary form　　　　　入境报关总表　　　　　第1章3.1节

enzyme [ˈenzaɪm] washing　　酵素洗　　　　　　　　第5章1.1.11节

Eton [ˈiːtn] collar　　　　　伊顿领　　　　　　　　第2章1.1.3节

European [ˌjuərəˈpi(ː)ne]
　　Article Number　　　　　欧洲物品编码　　　　　第3章3.2节

evening dress　　　　　　　晚礼服　　　　　　　　第1章1.1节

evening gown　　　　　　　（裙摆及地的女装）晚礼服　第1章1.3节

everyday clothing　　　　　日常衣着　　　　　　　第1章1节

expressly or tacitly　　　　明示或暗示地　　　　　第6章3.1节

extensibility [ɪksˌtensəˈbɪlɪtɪ]　可延伸性　　　　　第3章1.1节

eye-catching　　　　　　　吸引眼球的，引人注目的　第3章1.1节

eyelet　　　　　　　　　　气眼　　　　　　　　　第3章2.3节

eyelet [ˈaɪlɪt] buttonhole　　圆头纽孔（凤凰眼）　　第3章2.2.1节

eyelet buttonhole machine　　圆头锁眼机　　　　　　第3章2.2.1节

f

fabric area density　　　　　织物面密度　　　　　　第4章2.1节

fabric construction　　　　　织物结构　　　　　　　第1章2.1节

fabric fault　　　　　　　　织物疵点　　　　　　　第6章3.2.1节

fabric feeding system　　　　送料系统　　　　　　　第5章2.2.2节

fabric inspection　　　　　　织物检验　　　　　　　第5章1.1.4节

fabric inspection machine　　验布机　　　　　　　　第5章1.1.4节

fabric spreading　　　　　　铺料　　　　　　　　　第5章1.1.4节

fabric spreading machine　　铺料机　　　　　　　　第5章2.1.2节

fabric utilisation
　　[ˌjuːtɪlaɪˈzeɪʃən]　　　　织物利用率　　　　　　第2章1.1.2节

fabric width　　　　　　　　织物门幅　　　　　　　第5章1.1.4节

fabric-inspection machine　　验布机　　　　　　　　第5章2节

factory premise [ˈpremɪs]　　厂区内　　　　　　　　第5章1.1.13节

fashion sketch　　　　　　　时装效果图　　　　　　第5章1节

fashioned knitted garment　　成形编织的服装　　　　第4章2.3.4节

fastener [ˈfɑːsnə]　　　　　扣合件　　　　　　　　第3章2.2节

feed dog　　　　　　　　　送布牙　　　　　　　　第5章2.2.2节

feed wheel　　　　　　　　送料滚轮　　　　　　　第5章2.2.2节

felling machine　　　　　　缲边机　　　　　　　　第5章2.2.3节

femininity [femɪˈnɪnɪtɪ]　　女性气质　　　　　　　第1章1.1节

fictitious [fɪkˈtɪʃəs]　　　　虚构的　　　　　　　　第4章3节

filling [ˈfɪlɪŋ]　　　　　　　（絮状）填料　　　　　第3章2.4节

final pressing　　　　　　　成衣熨烫　　　　　　　第5章1.1.11节

finishing　　　　　　　　　整理　　　　　　　　　第5章1.1.11节

fish dart　　　　　　　　　鱼型省　　　　　　　　第2章1.3节

flame resistance	阻燃性	第6章3.1节
flame retardant [rɪˈtɑːdənt]	阻燃剂	第6章3.1节
flammability [ˌflæməˈbɪlətɪ]	易燃,可燃性	第3章3.1.5节
flared [fleəd] legs	喇叭裤脚	第1章1.2节
flat bed presser	平板式压烫机	第3章2.3节
flat collar	坦领	第1章1.1节
flat seam stitch	绷缝	第4章2.3.1节
flat seaming machine	绷缝机	第5章1.2节
flat weft knitting machines	平型纬编针织机,横机	第3章1节
flat-bed fusing press	平板式黏合机	第5章2.1.4节
flat-fold packaged [ˈpækɪdʒɪd]	折叠式包装的	第5章1.1.4节
flip [flɪp]	翻转	第2章2节
floor space	占地空间	第2章2节
floral [ˈflɔːrəl] pattern	花卉图案	第3章1.1节
fly [flaɪ]	裤门襟	第2章1.2.1节
fly length	裤门襟长	第2章3节
folding line	折边线	第2章2节
form finisher	人体模型整烫机	第5章2.3节
formal wear	正式衣装	第1章1.1节
formaldehyde [fɔːˈmældɪhaɪd]	甲醛	第5章1.1.4节
		第6章3.1节
foundation garment	(女)贴身内衣	第1章1.1节
frapped [fræpt] state	收紧的状态	第3章2.5节
fraying [freɪɪŋ]	散边	第3章2.1节
French curves	曲线板	第5章2节
French seam	法式缝(俗称"来去缝")	第4章2.3.4节
frill	褶边	第2章1.3节
front	前片	第2章1.1.1节
front facing [ˈfeɪsɪŋ]	前挂面	第2章1.1.1节
front neck drop	前颈深	第2章3节
front opening	前门襟开口	第3章2.2节
front opening placket	前门襟	第2章2节
front overlap	前叠门	第2章3节
front raglan	插肩袖前挂肩	第2章3节
front rise	前直裆	第2章3节
front sleeve	前袖片	第2章1.1.1节
front yoke depth	拼腰前深	第2章3节
fulling	缩绒	第5章1.2节
full-length sleeve	长袖	第2章1.1.2节
fully fashioned machine	全成形机	第5章1.2节
fully-fashioned [ˈfʊlɪ-ˈfæʃənd]	全成形的	第3章1节
fungi [ˈfʌŋgaɪ]	真菌	第6章3.1节
fur [fɜː] clothing	毛皮服装	第1章2.1节

fur lining	毛(皮)的夹里	第1章2.1节
fusible [ˈfjuːzəbl] interlining	热融衬	第3章2.3节
fusing	热融黏合	第2章1节
		第5章1.1.9节
fusing duration	上黏合衬的加热时间	第5章1.1.2节
fusing machine	黏合机	第5章2.1.4节
fusing temperature	上黏合衬的加热温度	第5章1.1.2节

g

gabardine [ˈgæbədiːn]/gaberdine	华达呢	第3章1.1节
garment bottom	下装	第1章1节
garment component	服装组成部分,衣片	第2章1节
garment length	衣长	第2章3节
garment making-up	服装成衣	第4章1节
garment order	服装订单	第2章2节
garment piece	衣片	第2章1节
garment section	服装部分,衣片	第1章2.3节
garment top	上装	第1章1节
garment wash	成衣洗水	第5章1.1.11节
gathered dart	带碎裥的省	第2章1.3节
gathers	碎裥	第2章1.3节
general rate	普通税率	第3章3.1.2节
gents' coat	男装风衣	第1章1.3节
glazed or yellowish spots	极光或焦斑	第6章3.2.2节
godet [gəuˈdet]	三角形布片	第2章1.2.2节
gore [gɔː]	三角形布片	第2章1.2.2节
gored [gɔːd] skirt	拼片裙	第1章1.2节
governmental legislation		
[ˌledʒɪsˈleɪʃən]	政府的立法	第6章2节
governmental legislature		
[ˈledʒɪsˌleɪtʃə]	政府的立法机构	第6章2节
governmental regulations	政府的规章	第6章2节
grading chart	推档表	第2章3节
grading rule	推档变量	第2章3节
grain [greɪn] line	丝缕线	第2章1.1.3节
gripper	拉柄	第3章2.2.2节
grown-on [grəun-ɒn]	连着的	第2章1.1.2节
grown-on collar	连身衣领	第2章1.1.3节
grown-on sleeve	连身袖	第2章1.1.2节

h

half circular skirt	(半圆裙片的)圆裙	第2章1.2.2节
half ring	半园扣	第3章2.5节

half slip	下装衬裙	第1章1.1节
halter [ˈhɔːltə] top	吊颈(上装)背心	第1章3.2节
hand painted kimono [kɪˈməʊnəʊ]	手绘和服	第5章1.1.11节
hand stitch	手工线迹	第4章2.3.1节
handle	手感	第5章1.1.9节
handling system	搬运系统	第5章2.2.4节
hand-stitch machine	仿手工线迹缝纫机	第4章2.3.1节
hanger [ˈhæŋə]	衣架	第5章1.1.13节
hanger hook	衣架挂钩	第5章1.1.13节
hangtag [ˈhæŋtæg]	吊牌	第3章3节
hank [hæŋk]	纱绞	第5章1.2节
hard and fast rule	一成不变的法则	第4章2节
harmful contaminant [kənˈtæmɪnənt]	有害的污染物	第6章3.1节
Harmonized [ˈhɑːmənaɪzd] Commodity Description and Coding System (H. S. Code)	协调制商品名称和编码方法	第1章2.1节
Harmonized Tariff Schedule	美国协调关税制	第1章3.1节
Hawaiian [hɑːˈwaɪɪən] shirt	夏威夷衫	第1章1.3节
Hawaiian dress	夏威夷裙	第1章3.2节
heat shrinkage [ˈʃrɪŋkɪdʒ]	热收缩	第5章1.1.2节
heating element	加热元件	第5章2.1.4节
heating section	加热区	第5章2.1.4节
heat-safe padding	耐热衬垫	第5章2.3节
heavy metal	重金属	第5章1.1.4节
heavy weight fabric	厚型织物	第3章1.1节
heavy-duty textile	厚实的纺织物	第6章3.1节
hem	下摆	第2章3节
hem height	下摆宽(折边高)	第2章3节
hemming [ˈhemɪŋ]	缲边	第4章2.3.1节
hemming attachment	缲边装置	第5章1.1.10节
high point shoulder (HPS)	高肩点	第2章3节
high waist	高腰	第1章1.2节
higher shoulder point (HSP)	高肩点	第2章3节
hinged [hɪndʒd] presser foot with two toes	双趾铰式压脚	第5章2.2.2节
hip pocket	臀后的袋	第2章1.1.4节
hip section [ˈsekʃən]	裙或裤的袋垫	第2章1.2.1节
hole	破洞	第6章3.2.1节
hood [hʊd]	风帽	第2章1.1.1节
hood crown length	帽顶长	第2章3节
hood height	帽高	第2章3节
hood opening length	帽开口长	第2章3节

hood width	帽宽	第2章3节
hook point	梭钩尖	第5章2.2.2节
hormone ['hɔːməʊn]	荷尔蒙,激素	第6章3.1节
humidity [hjuːˈmɪdɪtɪ]	湿度	第5章1.1.5节
hybrid system	混合作业法	第5章2.2.4节
hydraulically [haɪˈdrɔːlɪkəlɪ]		
powered equipment	液压传动设备	第5章2.1.3节

i

icon ['aɪkɒn]	图标	第5章1.1.5节
illegal [ɪˈliːgəl] entrepot		
['ɒntrəpəʊ] trade	非法转口贸易	第3章3.1.2节
imitation [ɪmɪˈteɪʃən] silk lapel	仿丝质驳领	第1章2.1节
immune [ɪˈmjuːn]	免疫的	第6章3.1节
imperfection	缺陷	第5章1.2节
in short interval ['ɪntəvəl]	短片段的	第5章1.1.4节
incontinent [ɪnˈkɒntɪnənt]	失禁的	第3章1节
inherent [ɪnˈhɪərənt] quality	内在质量	第5章1.1.4节
inner pocket	内袋	第2章1.1.4节
input	输入	第2章2节
inseam ['ɪnsiːm]	裤脚长,裤脚内侧缝	第2章3节
insert [ɪnˈsɜːt] pocket	挖袋,镶嵌袋	第2章1.1.4节
insertion pin	插针	第3章2.2.2节
intellectual [ˌɪntəˈlektjʊəl]		
property ['prɒpətɪ]	知识产权	第3章3.1.1节
interconnect	相互连接	第4章2.3.1节
interlacing [ˌɪntə(ː)ˈleɪsɪŋ]	线环(圈)交叉	第4章2.3.1节
interlining [ˌɪntəˈlaɪnɪŋ]	衬头	第2章1.1.2节
interlock [ˌɪntəˈlɒk]	棉毛布(双罗纹)	第3章1.1节
interlooping [ˌɪntə(ː)ˈluːpɪŋ]	线环(圈)互串	第4章2.3.1节
intermediate and final inspections	中期检验和最终检验	第4章2.4节
international size label	国际尺码标签	第3章3.1.3节
inverted pleat	阴裥	第2章1.3节
invisible zipper	隐形拉链	第3章2.2.2节
ironing ['aɪənɪŋ] table	烫台	第5章2.3节
ironing mark	烫痕	第6章3.2.2节
irregular stitch	线迹不匀	第6章3.2.2节
isotropic [aɪsəʊˈtrɒpɪk]	各向同性	第3章2.3节

j

jacket ['dʒækɪt]	夹克衫,短上装外衣的统称	第1章1.1节
jacket press	上衣压烫机	第5章2.3节
jacquard [dʒəˈkɑːd]	提花的	第3章3.1.1节

jeans [dʒiːnz] 　　　　　　　牛仔裤 　　　　　　第1章1.2节
　　　　　　　　　　　　　　　　　　　　　　　第5章1.1.11节

jig sewing 　　　　　　　　　模板缝纫 　　　　　　第5章2.2.3节
jogging ['dʒɒgɪŋ] bottoms 　慢跑时穿的运动裤 　第1章1.2节
joining 　　　　　　　　　　缝合 　　　　　　　　第3章2.1节
jumper 　　　　　　　　　　（英）（粗）针织套衫 第1章1.1节
jumper/ jumper dress 　　　（美）无袖套头连衣裙 第1章1.1节
jumpsuit 　　　　　　　　　伞兵装,连裤外衣 　　第1章1.3节

k

keyboard 　　　　　　　　　键盘 　　　　　　　　第5章2.1.2节
khaki ['kɑːkɪ] 　　　　　　卡叽 　　　　　　　　第3章1.1节
kilt [kɪlt] 　　　　　　　　苏格兰格子呢裙 　　第1章1.2节
knee（girth） 　　　　　　　膝围(围长) 　　　　第2章3节
knickers ['nɪkəz] 　　　　　（英）（女）短内裤 第1章1.2节
knickers/ knickerbockers
　['nɪkəbɒkəz] 　　　　　　灯笼裤 　　　　　　第1章1.2节
knife pleat 　　　　　　　　刀裥 　　　　　　　第2章1.3节
knit-fashioned 　　　　　　编织成形 　　　　　第1章2.3节
knitted fleece fabric 　　　针织起绒布 　　　　第1章1.1节
knitted garment 　　　　　针织服装 　　　　　第1章1.1节
knitted rib cuff ['kʌf] 　　针织罗纹袖口 　　第2章1.1.1节

knitted rib waistband
　['weɪstbænd] 　　　　　　针织罗纹腰头 　　第2章1.1.1节
knitted shirt 　　　　　　　针织翻领衫 　　　第1章3.2节
knitted suit 　　　　　　　针织套装 　　　　　第1章1.3节
knitwear ['nɪtˌweə] 　　　针织服装 　　　　　第1章3.2节
knot [nɒt] 　　　　　　　　打节 　　　　　　　第3章2.1节

l

label ['leɪbl] 　　　　　　　标签 　　　　　　　第3章3节
labeling gun 　　　　　　　打标签贴的枪 　　　第5章2.1.5节
lace 　　　　　　　　　　　花边 　　　　　　　第1章1.1节
ladies' coat 　　　　　　　女装风衣 　　　　　第1章1.3节
lapel [lə'pel] 　　　　　　驳领 　　　　　　　第1章1.1节
lapped seam 　　　　　　　搭接缝 　　　　　　第4章2.3.3节
lay 　　　　　　　　　　　铺好的布层 　　　　第5章1.1.5节
lay planning 　　　　　　　铺料设计 　　　　　第5章1.1.4节
leather ['leðə] 　　　　　　皮革 　　　　　　　第3章1.1节
leather-imitation button 　仿皮扣 　　　　　　第3章2.2.1节
left back 　　　　　　　　左后片 　　　　　　第2章1.1.1节
left back side 　　　　　　左后侧片 　　　　　第2章1.1.1节

left front	左前片	第2章1.1.1节
left front side	左前侧片	第2章1.1.1节
leg cuff	裤脚口	第2章1.2.1节
leg opening	裤脚口大	第2章3节
leg pocket	裤腿上的裤袋	第2章1.1.4节
leisure ['leʒə; 'li:ʒə] wear	休闲服	第1章1.2节
levy ['levɪ]	征收	第1章2.3节
light weight fabric	薄型织物	第3章1.1节
ligne [li:n]	(纽扣)号数	第3章2.2.1节
linear ['lɪnɪə] density	线密度	第5章1.2节
		第3章2.1节
lined garment	有夹里的服装	第2章2节
lingerie ['lænʒəri:]	(女)内衣	第1章1.1节
lining	夹里,里子	第2章1.1.1节
lining	里料	第2章2节
linking machine	套口机	第5章1.2节
literal ['lɪtərəl] instructions	文字说明	第3章3.2节
liver ['lɪvə]	肝脏	第6章3.1节
loading station	上料台	第5章2.1.4节
lock stitching machine	锁式缝机(俗称"平缝机"、"平车")	第5章1.2节
locknit ['lɒknɪt]	(经编)经平绒	第3章1.2节
lockstitch ['lɒkstɪtʃ]	锁式线迹	第4章2.3.1节
lockstitch machine	锁式缝缝纫机(平缝机)	第5章2.2.1节
long groove	长针槽	第5章2.2.2节
long or short sleeved	长袖或短袖的	第1章1.1节
long-sleeved vest	长袖内衣	第1章1.1节
loop density	线圈密度	第5章1.2节
loop transfer	移圈	第5章1.2节
loop/stitch structure	线圈结构	第1章1.1节
looper	弯针	第5章2.2.2节
looper ['lu:pə] thread	弯针线	第3章2.1节
loose-fitting	宽松的	第1章1.1节
loose-fitting garment	宽松式服装	第4章2.4节
lounge [laʊndʒ] suit	普通西服套装	第1章1.1节
lower front	下前片	第2章1.1.1节
lower garment	下装	第1章1节
lower hip (girth)	臀围(围长)	第2章3节
Lutrabond ['lu:trəbɒnd]	(卢特拉邦德品牌)纸膜	第3章2.4节
luxurious [lʌg'zjʊərɪəs]	奢华的,豪华的	第3章1.1节
Lycra ['laɪkrə]	莱卡(弹力纤维)	第3章1.1节

m

machine gauge [geɪdʒ]	机号	第5章1.2节
machinists	缝纫工	第5章2.2.4节
Mackintosh ['mækɪntɒʃ]	胶布雨衣	第1章1.3节
magnetic [mæg'netɪk] button	磁性扣	第3章2.5节
to make out...	缮制……	第5章1.1.13节
mandarin ['mændərɪn] collar	中式立领	第2章1.1.3节
mandatory ['mændətərɪ] inspection	强制检验	第3章3.1.2节
man-made fibre	化学纤维	第1章2.2节
man-made material	化纤材料	第3章1.1节
manufacturing sequences	生产流程	第5章1节
marble ['mɑːbl] grain [greɪn]	大理石纹理的	第3章2.2.1节
marine [mə'riːn]	海军蓝	第5章1.1.5节
marker ['mɑːkə]	(衣片样板)排料图,排板图	第5章1.1.5节
		第2章1.1.2节
marker making	(衣片样板)排板,排料	第2章1.1.2节
		第5章1.1.4节
masculine ['mɑːskjʊlɪn] look	阳刚气的外表	第2章1.3节
material and labour costs	材料及人工成本	第5章1.1.2节
measurement	测量,测量尺寸	第2章3节
measurement chart	尺寸表	第2章3节
		第4章2节
measurement location	测量部位	第2章3节
measurement tolerance ['tɒlərəns]	测量容差	第4章2.4节
mechanical property	机械性能	第5章1.1.4节
mechanical strength	机械强度	第6章3.1节
medium ['miːdjəm] size	中档尺码	第4章2.1节
melon ['melən] sleeve	瓜型袖	第2章1.1.2节
men's coat	男装风衣	第1章1.3节
metal claw [klɔː]	金属爪扣	第3章2.2.2节
metal zipper	金属拉链	第3章2.2.2节
metal-coated	镀金属的	第3章2节
metre rule	米尺	第5章2节
metric ['metrɪk] count	公制支数	第3章2.1节
mid-calf [mɪd-kɑːf] length	至腿肚长的	第2章1.2.1节
milling	缩绒	第5章1.2节
mini ['mɪnɪ] skirt	超短裙	第1章1.2节
miniaturised ['mɪnɪətʃəraɪzd]	缩小的	第5章1.1.5节
monofilament ['mɒnəʊ'fɪləmənt] thead	单丝缝纫线	第3章2.1节
morning dress	常礼服	第1章2.1节
morning/wedding suit	常礼服套装/婚礼服套装	第1章1.3节

mother of pearl	贝母	第6章3.1节
moth-proof finishing	防蛀整理	第5章1.2节
motif [məuˈtiːf]	图案	第5章1.1.5节
mould [məuld]	发霉	第6章3.1节
multifilament [ˌmʌltɪˈfɪləmənt]	多孔丝，复丝	第3章2.1节
multiple continuous filaments	多孔长丝	第3章1.2节

n

nape（centre back neck）	后颈点（领圈背中线处）	第2章3节
（the length）from nape [neɪp]		
to waist	背长	第2章3节
national dress	民族服装	第1章1.1节
natural [ˈnætʃərəl] fibre	天然纤维	第3章1.1节
		第5章1.1.4节
neaten	（毛边）收光	第5章1.1.10节
neatening [niːtnɪŋ]	毛边收光，毛边处理	第2章1.3节
neck circumference [səˈkʌmfərəns]	颈围	第2章3节
neck dart	领圈处的省	第2章1.3节
neck depth	领深	第2章3节
neckline [ˈneklaɪn]	领围线，领圈	第2章1.1.3节
needle blade	针身（针杆）	第5章2.2.2节
needle eye	针眼	第5章2.2.2节
needle point	针尖	第5章2.2.2节
needle position motor	缝针定位马达	第5章2.2.3节
needle thread	缝针线	第4章2.3.1节
needling system	刺料系统	第5章2.2.2节
nervous [ˈnɜːvəs]	神经的	第6章3.1节
net width	净门幅	第5章1.1.5节
neutral [ˈnjuːtrəl] packing	中性包装	第3章3.1.2节
nightwear [ˈnaɪtweə]	睡衣	第3章3.1.5节
Nomex [ˈnəumeks]	诺梅克斯（一种耐高温芳香族聚酰胺，杜邦公司品牌）	第3章2.1节
non-fraying [nɒn-freɪŋ] fabric	不散边织物	第4章2.3.4节
non-irritating [nɒn-ˈɪrɪteɪtɪŋ]	不会过敏的	第3章1.1节
non-woven fabric	非织造织物	第3章1节
notch	刀眼	第2章1.3节
numbering [ˈnʌmbərɪŋ] gun	打号枪	第5章2.1.5节
nylon [ˈnaɪlən]	尼龙	第3章1.1节
nylon bristle [ˈbrɪsl]	尼龙刺须	第5章2.1.3节
nylon taffeta	尼丝纺	第3章1.2节
nylon zipper	尼龙拉链	第3章2.2.2节

o

OEKO-TEX standard	OEKO-TEX 标准	第3章3.1.5节
off-grain	偏离丝缕线	第5章1.1.5节
office-wear	办公室着装	第1章1.3节
official language	官方语言	第3章3.2节
oil spot	油污迹	第6章3.2.1节
oiling [ˈɒɪlɪŋ]	上油	第5章1.2节
one-piece sleeve	一片袖	第2章1.1.2节
one-way pile	单向绒毛	第5章1.1.5节
open width	开幅的	第5章1.1.6节
open zipper	开尾拉链	第3章2.2.2节
origin [ˈɒrɪdʒɪn] label	原产地标签	第3章3.1节
original sample	原样	第5章1.1.12节
originality [əˌrɪdʒɪˈnælɪtɪ]	创意	第5章1节
outerwear [ˈaʊtəweə(r)]	外衣	第1章1节
outseam [ˈaʊtˌsiːm]	裤长	第2章3节
Outward Processing Program	外加工	第1章3.1节
overalls [ˈəʊvərɔːlz]	工装裤	第1章1.2节
overcoat	(厚)大衣	第1章1.1节
overedge [ˈəʊvəˌedʒ] stitching	包缝	第3章2.1节
overedge seaming	包缝,包边缝	第4章2.3.1节
overhead,computerized and me-chanized chain handling system	架空的、计算机控制的、机械式的链条式搬运系统	第5章1.1.8节
overlap [ˈəʊvəˈlæp] (width)	叠门(宽)	第2章3节
overlock [ˈəʊvəlɒk] stitch	包缝线迹	第3章2.1节
overlock machine	包缝机	第5章2.2.1节
overlock stitch	包缝	第4章2.3.1节
overlocking machine	包缝机	第5章1.2节
overpressing	过分熨烫	第6章3.2.2节
oversewn	拷边,包缝	第4章2.3.4节
overshirt	宽松式可穿在针织衫等外的衬衣	第1章1.3节

p

packing list	装箱单	第5章1.1.13节
to be paid by piece work	计件付酬	第5章2.2.4节
palace [ˈpælɪs]	派力司	第3章1.1节
panel [ˈpænl]	(条形)衣片	第2章1.3节
panties [ˈpæntɪz]	(美)(女)短内裤	第1章1.2节
pants	(美)外衣长裤,(英)长内裤	第1章1.2节
panty-hose	连裤袜	第1章1.2节
paper pattern	纸样	第2章2节
parka [ˈpɑːkə]	派克大衣,长夹克式休闲大衣	第1章1.1节
patch [pætʃ] pocket	贴袋	第2章1.1.4节

pattern ['pætən]	衣片样板	第2章2节
pattern cutter	样板设计师,打板人员	第5章1.1.1节
pattern cutting	样板制作,出服装样板,打板	第5章1.1.1节
pattern master	打板专用尺	第5章2节
pattern matching	对花	第5章1.1.5节
pattern notches	(样板)刀眼夹钳	第5章2节
pattern number	样板编号	第4章2节
PCB（polychlorinated [ˌpɒliˈklɔːrɪneɪtɪd] biphenyl [baɪˈfiːnɪl]）	多氯联苯	第6章3.1节
PCP（Pentachlorophenol [ˈpentəˌklɔːrəˈfiːnɒl]）	五氯苯酚	第6章3.1节
PCT（polychlorinated terphenyl [təˈfiːnɪl]）	多氯三联苯	第6章3.1节
pearlescent [pɜːˈlesnt]	珠光的	第3章2.2.1节
peg [peg] top skirt	上宽下窄的裙子	第1章1.2节
pendant [ˈpendənt]	坠子	第3章2.2.2节
penetration [peniˈtreɪʃən]	穿透	第4章2.3.3节
perforated paper	打孔纸	第5章2.1.3节
performance	绩效	第5章2.2.4节
personality [ˌpɜːsəˈnælɪti]	个性	第1章1.3节
pesticide [ˈpestɪsaɪd]	杀虫剂	第5章1.1.4节
Peter Pan collar	彼得·潘领	第2章1.1.3节
piece by piece	逐匹地	第5章1.1.4节
piece dyed fabric	匹染的织物	第5章1.1.4节
piece length	匹长	第5章1.1.4节
piece name	衣片名称	第2章2节
pig nose toggle [ˈtɒgl]	椭圆型双眼绳塞	第3章2.5节
pile direction	绒毛方向	第5章1.1.5节
pill-resistant finishing	防起球整理	第5章1.2节
pin lock	拉链头锁钩	第3章2.2.2节
pinafore [ˈpɪnəfɔː] dress	(英)无袖套头连衣裙	第1章1.1节
pinking	锯齿型布边裁剪	第4章2.3.4节
pinking [pɪŋkɪŋ] shear	锯齿形布边剪	第4章2.3.4节
piping [ˈpaɪpɪŋ]	滚边,嵌线	第1章2.1节
		第2章1.1.4节
piping [ˈpaɪpɪŋ] pocket	嵌线袋	第2章1.1.4节
piping foot	嵌线压脚	第5章2.2.2节
placket [ˈplækɪt]	门襟	第2章1.1.1节
plain	素色的,普通的	第1章2.1节
plain [pleɪn] weft [weft] knit	纬编平针	第3章1.1节
plain woven structure	机织平纹织物	第3章1.1节
plastic [ˈplæstɪk, ˈplɑːstɪk] zipper	塑齿拉链	第3章2.2.2节

plastic sleeve	塑料套管	第3章2.5节
plastisol [ˈplæstɪsɒl] prints	增塑溶胶印花	第6章3.1节
pleat [pliːt]	裥	第2章1.2.2节
pleated [pliːtɪd]	打褶的,打裥的	第2章1.2.1节
pleated skirt	褶裥裙	第1章1.2节
plotter	绘图机	第5章2节
plotting	(绘图机)绘制	第5章1.1.5节
ply slippage [ˈslɪpɪdʒ]	缝料滑移	第5章2.2.2节
pneumatically [nju(ː)ˈmætɪkəlɪ]		
powered equipment	气动设备	第5章2.1.3节
pocket	衣袋	第1章1.1节
pocket [ˈpɒkɪt] opening	袋口	第2章1.1.4节
pocket bag	袋布,衣袋	第2章1.1.4节
pocket flap [flæp]	袋盖	第2章1.1.4节
the point of rotating hook	旋梭梭尖	第5章2.2.2节
polo [ˈpəʊləʊ] shirt	套头式针织运动衫	第1章1.1节
polybag	塑料袋	第5章1.1.13节
polyester [ˌpɒlɪˈestə]	涤纶	第3章1.1节
polyester pongee [pɒnˈdʒiː]	春亚纺	第3章1.2节
polyester taffeta	涤丝纺	第3章1.2节
polyester wadding [ˈwɒdɪŋ]	涤纶喷胶棉	第3章1.2节
poplin [ˈpɒplɪn]	府绸	第3章1.1节
porosity [pɔːˈrɒsɪtɪ]	多孔性,透气性	第3章1.1节
press button	揿纽,铐纽	第3章2.2.3节
presser [ˈpresə] foot	压脚	第5章2.2.2节
pressing roller	轧辊	第5章2.1.4节
princess line	公主缝	第2章1.1.1节
printed,knitted or woven pattern	印花、针织或机织的图案	第5章1.1.5节
printer	打印机	第5章2节
production line	产生线,流水线	第5章2.2.4节
progressive bundle system	逐捆作业方法	第5章2.2.4节
prong [prɒŋ]	戳销,爪扣	第3章2.2.2节
propensity [prəˈpensɪtɪ] to		
pilling [ˈpɪlɪŋ]	起球的倾向	第3章1.1节
property [ˈprɒpətɪ] (right)	知识产权	第3章3.1.1节
puckering	缝迹起皱	第4章2.3.3节
puff [pʌf] sleeve	泡泡袖	第2章1.1.2节
pullover [ˈpʊləʊvə(r)]	套衫	第1章1.1节
pull-tab	拉柄	第3章2.2.2节
purchase [ˈpɜːtʃəs] contract	购货合同	第5章1.1.4节
PVC and PU fabric	聚氯乙烯和聚氨酯织物	第6章3.1节

q

quilting ['kwɪltɪŋ]	绗缝	第3章2.1节
quilting machine	绗缝机	第5章2.2.3节
quotation [kwəʊ'teɪʃən]	报价	第3章2.1节

r

racer ['reɪsə] back vest	窄背背心	第1章3.2节
raglan ['ræglən] sleeve	插肩袖	第2章1.1.1节
rain coat	雨衣	第1章1.1节
random sampling	随机取样	第5章1.1.12节
randomly ['rændəmlɪ]	随机地	第5章1.1.4节
raw [rɔː] edge	毛边	第3章2.1节
rayon ['reɪɒn] satin ['sætɪn]	(人造丝)光缎羽纱	第3章1.2节
real time production monitoring ['mɒnɪtərɪŋ]	实时生产监控	第5章2.2.4节
reciprocal [rɪ'sɪprəkəl] feeding	往复送料	第5章2.2.2节
rectify ['rektɪfaɪ]	调整	第5章1.1.12节
regional ['riːdʒənəl] regulations	区域性规则	第1章3.1节
relax	松弛	第5章1.2节
remedy ['remɪdɪ]	补救	第5章1.1.12节
residual [rɪ'zɪdjʊəl]	滞留的,残余的	第5章1.1.4节
residual	残留物	第6章3.1节
resin ['rezɪn]	树脂	第5章1.1.9节
resultant [rɪ'zʌltənt] cotton count	总英支数	第3章2.1节
retaining [rɪ'teɪnɪŋ] box	(拉链)针盒	第3章2.2.2节
rever [rɪ'vɪə]	驳领	第2章1.1.3节
reverse [rɪ'vɜːs] locknit	(经编)经绒平	第3章1.2节
(1×1 or 2×2) ribs	(1+1 或 2+2)罗纹	第3章1.1节
right back	右后片	第2章1.1.1节
right back side	右后侧片	第2章1.1.1节
right front	右前片	第2章1.1.1节
right front side	右前侧片	第2章1.1.1节
right side (RS)	织物正面	第2章1.1.2节
rivet ['rɪvɪt]	铆钉,工字扣	第3章2.3节
roll collar	翻领	第2章1.1.3节
rotary cutter	圆刀型(电)裁刀	第5章2.1.3节
rotating hook	旋梭	第5章2.2.2节
rotating looper	旋转钩针	第5章2.2.2节
rouleau [ruː'ləʊ]	滚条	第2章1.1.4节
round knife cutter (round knife)	圆刀型(电)裁刀	第5章2.1.3节
round neck	圆领	第1章1.1节
round point	球形针尖	第5章2.2.2节
rounded tails	圆弧后摆	第1章2.1节

S

safety stitch	安全缝	第4章2.3.1节
sample	样品	第1章3.2节
sample making	打样	第3章2.1节
sample pattern	样衣的样板	第5章1.1.3节
sample room	打样间	第5章1.1.2节
sample-making	打样	第5章1.1.2节
sand washing	砂洗	第5章1.1.11节
sandwich	夹(着)缝(上)	第5章1.1.10节
saree ['sɑːrɪ]	纱丽	第1章1.2节
sarong ['sɒːrɒŋ,səˈrɒŋ]	纱笼	第1章1.2节
scanner	扫描仪	第5章2节
scarf	针缺口	第5章2.2.2节
scraps [skræps] of fabrics	织物碎片,碎料	第2章1.3节
seam	缝子	第2章1.3节
seam allowance [əˈlaʊəns]	缝头	第2章2节
seam neatening	缝边收光	第4章2.3节
seam pucker	缝迹起皱	第6章3.2.2节
seam type	缝子类型,缝型,缝式	第4章2.3节
seaming	缝合	第4章2.3.1节
seamless ['siːmlɪs]	无缝的	第3章1节
second-hand	二手的,旧的	第1章2.1节
self-neatening	缝子自行光边	第4章2.3.4节
selvedge ['selvɪdʒ]	布边	第5章1.1.4节
sensor	传感器	第5章2.2.3节
separates	上下装分开销售的服装	第1章1.3节
serge [səːdʒ]	哔叽	第3章1.1节
set	套	第1章1.3节
set-in sleeve	装袖	第2章1.1.2节
sewing	缝纫	第2章1节
sewing fault	缝纫疵点	第5章1.1.10节
sewing machine	缝纫机	第5章2.2节
sewing threads	缝纫线	第3章2.1节
sewn-in lining	缝上去的夹里	第3章1.2节
shank	针柄	第5章2.2.2节
shank [ʃæŋk] button	有脚扣	第3章2.2.1节
shape retention [rɪˈtenʃən]	保型性	第3章1.1节
shaped panel	成形的衣片	第3章1节
shawl [ʃɔːl] collar	青果领,披肩领	第2章1.1.3节
shears	剪刀	第5章2节
shell [ʃel]	面料	第1章2.1节
shirt [ʃɜːt]	衬衫	第1章1.1节
shirt press	衬衫压烫机	第5章2.3节

shoe lace	鞋带	第3章2.5节
short groove	短针槽	第5章2.2.2节
short sleeve	短袖	第2章1.1.2节
shorts	短裤	第1章1.2节
short-sleeved vest	短袖内衣	第1章1.1节
shoulder	针肩	第5章2.2.2节
shoulder dart	肩部的省	第2章1.3节
shoulder neck point	肩颈点	第2章3节
shoulder pad	肩衬	第3章2.5节
shoulder point	肩点	第2章3节
shoulder slope	肩斜	第2章3节
shoulder strap [stræp]	开衩处的镶片	第2章1.3节
shrinkage ['ʃrɪŋkɪdʒ]	缩水(率),收缩	第3章1.1节
		第5章1.1.4节
shrink-proof finishing	防缩整理	第5章1.2节
side seam	(上衣)摆缝,(裤或裙)侧缝, (裤)栋缝	第2章1.1.1节
side yoke depth	拼腰侧深	第2章3节
significant level	显著性等级	第6章3.2.1节
silhouette [ˌsɪlu(ː)'et]	轮廓,侧影,造型	第1章1.2节
silhouette [ˌsɪlu(ː)'et]	轮廓,侧影,造型	第1章1.2节
silk blends	真丝混纺	第1章2.2节
silk habotai	真丝电力纺	第3章1.1节
single breasted ['brestɪd]	单排纽	第1章1.1节
single jersey ['dʒɜːzɪ]	单面针织物	第3章1.1节
single piping foot	单嵌线压脚	第5章2.2.2节
single thread chain stitch	单线链缝	第4章2.3.1节
single-needle lockstitch machine	单针平缝机	第5章2.2.1节
singlet	背心	第1章1.1节
singlet ['sɪŋglɪt] top	(女)吊带衫	第1章1节
size chart	尺寸表	第2章3节
size label	尺码标签	第3章3.1节
size specifications	尺寸表	第2章3节
skin cancer	皮肤癌	第6章3.1节
skipped stitch	跳线	第6章3.2.2节
skirt	裙	第1章1.2节
skirt block	裙的基样	第5章1.1.1节
skirt length	裙长	第2章3节
skirt suit	裙套	第1章3.2节
slacks	宽松裤	第1章1.2节
sleeve block	袖的基样	第5章1.1.1节
sleeve crown [kraʊn]	袖山	第2章1.1.2节
sleeve hem [hem]	袖口边	第2章1.1.2节

sleeve length	袖长	第2章3节
sleeve opening	袖口大	第2章3节
sleeve strap	袖襻	第2章1.3节
sleeveless	无袖的	第1章1.1节
slider ['slaɪdə]	(带上的)滑扣,无针带夹	第3章2.5节
slider body	拉链头座	第3章2.2.2节
slip	衬裙	第1章1.1节
slit [slɪt]	开衩	第2章1.2.2节
slit facing	开衩处贴边	第2章1.3节
slit strip [strɪp]	开衩处的滚条	第2章1.3节
slub catcher	清纱器	第5章1.2节
snap [snæp]	撤纽,铐纽	第3章2.2.3节
snarling ['snɑːlɪŋ]	缠结	第3章2.1节
snorkel ['snɔːkl] parkas	(风帽小开口的)派克衫	第1章1.1节
snug [snʌg]	非常贴身的,舒适的	第3章1.1节
socket ['sɒkɪt]	撤纽阴扣	第3章2.2.3节
softener ['sɒfnə]	柔软剂	第6章3.1节
software ['sɒftweə]	软件	第2章2节
solid	素色的,单色的	第5章1.1.5节
spaghetti [spə'getɪ] singlet top	多带式吊带衫	第1章3.2节
Spandex ['spændeks]	斯潘德克斯弹力纤维	第3章1.1节
spare [speə] button	备用扣	第3章2.2.1节
special rate	特别税率	第3章3.1.2节
specification sheet (SPEC sheet)	工艺单	第4章1节
specifications	规格,规格说明	第4章2.1节
spiral ['spaɪərəl] zipper	尼龙拉链	第3章2.2.2节
spline [splaɪn] function	样条函数	第5章1.1.3节
sportswear	运动服	第1章1.3节
spreader	线叉	第5章2.2.2节
spring loaded cord stop	弹簧型绳塞	第3章2.5节
square [skweə] neck	方领(圈)	第2章1.1.3节
stacking ['stækɪŋ] device	堆布装置	第5章2.2.3节
stain removing machine	去污迹机	第5章2.3节
standard minute value	标准测时值	第5章2.2.4节
staple ['steɪpl] spun [spʌn]	短纤维纺制的	第3章2.1节
statistical [stə'tɪstɪkəl] data	统计数据	第1章3.1节
statistics [stə'tɪstɪks]	统计学,统计表	第1章3.1节
stay tape	接缝狭带,牵条	第6章3.2.2节
steam iron ['aɪən]	蒸汽熨斗	第3章2.3节
steam press	蒸汽压烫机	第5章1.2节
steaming machine	蒸汽定型机	第5章1.2节
stereotyped ['stɪərɪətaɪpt]	固定套路的,老一套的	第1章1.1节
stiffness ['stɪfnɪs]	硬挺性	第3章1.1节

stitch against stitch	线圈对线圈	第5章1.2节
stitch cycle	线迹形成循环,成缝周期	第4章2.3.2节
stitch density	线迹密度	第4章2.3节
stitch formation cycle	成缝循环	第5章2.2.2节
stitch forming system	成缝系统	第5章2.2.2节
stitch type	线迹类型	第4章2.3节
stitching	缝合	第2章1.3节
stocking [ˈstɒkɪŋ]	长袜	第1章3.2节
stonewashed	经石洗的	第1章1.2节
stone washing	石洗	第5章1.1.11节
straight buttonhole	平头纽孔	第3章2.2.1节
straight buttonhole machine	平头锁眼机	第3章2.2.1节
		第5章2.2.3节
straight knife cutter （straight knife）	直刀型(电)裁刀	第5章2.1.3节
straight legs	直筒裤脚	第1章1.2节
straight skirt	直筒裙	第1章1.2节
strain	应变	第5章1.2节
streak	条痕	第6章3.2.1节
stress	应力	第5章1.2节
strike back	黏合剂渗穿织物	第5章1.1.9节
strike through	黏合剂渗穿织物	第5章1.1.9节
string channel [ˈtʃænl]	绳槽	第3章2.5节
string tunnel [ˈtʌnl]	绳槽	第3章2.5节
strip-cutting machine	开滚条机	第5章2.1.5节
stripe [straɪp] or check design	条纹或格子花型	第2章1.1.4节
striped [straɪpt] trousers	条纹长裤	第1章2.1节
strip-off device	剥离装置	第5章2.1.4节
stroke [strəʊk]	工作动程	第5章1.1.5节
structural [ˈstrʌktʃərəl] pocket	结构袋	第2章1.1.4节
structural line	结构线	第5章1.1.1节
stud [stʌd]	撳纽阳扣	第3章2.2.3节
style	款式	第1章1.1节
style number	款号	第2章2节
		第3章2.2.1节
		第4章2节
styling feature	款式特征	第2章1节
sub-assembly operation	次级的成衣操作,成衣 中间的操作	第5章2.2.4节
subcontract [sʌbˈkɒntrækt]	合同转包,外包	第5章1.1.11节
sub-contracted [sʌb-kənˈtræktɪd]	(工程)外包	第3章2.2.1节
substitute [ˈsʌbstɪtjuːt] button	备用扣	第3章2.2.1节
substrate [ˈsʌbstreɪt]	底布	第3章2.3节

suit [sjuːt]	套装,西服套装	第1章1.1节
sundress ['sʌnˌdres]	(露肩)背带式连衣裙	第1章1.1节
sunray-pleated skirt	百褶裙	第1章3.2节
superimposed seam	叠缝	第4章2.3.3节
surgical ['sɜːdʒɪkəl] gown	手术袍	第3章1节
swatch [swɒtʃ]	样布	第5章1.1.4节
sweater ['swetə]	(粗)针织衫,针织毛衣	第1章1.1节
sweatshirt ['swetʃɜːt]	长袖运动衫	第1章1.1节
sweep [swiːp]	下摆围长	第2章1.2.2节
swimming trunks ['trʌŋks]	(男)泳裤	第1章1.2节
swimwear [swɪmweə]	泳装	第3章2.1节
symmetrical [sɪ'metrɪkəl]	对称的	第2章2节
synthetic [sɪn'θetɪk] material	合成纤维材料	第3章1.1节

t

t. i. t. (tone in tone)	配色	第4章2.1节
taffeta ['tæfɪtə]	塔夫绸	第3章1.1节
tailcoat	燕尾服	第1章2.1节
tailor's chalk	划粉	第5章1.1.5节
tailored ['teɪləd] garment	度身裁制的服装	第1章1.1节
take-up cam	挑线凸轮	第5章2.2.2节
take-up lever	挑线杆	第5章2.2.2节
tank top	无袖圆领衫	第1章1.1节
tape measure	软尺,皮带尺	第5章2节
tapered ['teɪpəd] legs	小裤脚	第1章1.2节
tearing strength	撕裂强度	第6章3.1节
technical parameter	技术参数	第4章1节
TeCP (2,3,5,6-Tetrachloro-phenol [ˌtetrəˌklɔːrəˈfiːnɒl])	2,3,5,6-四氯苯酚	第6章3.1节
template ['templɪt]	模板	第2章2节 第5章1.1.10节
template sewing	模板缝纫	第5章2.2.3节
tenacity [tɪ'næsɪtɪ]	强度	第3章2.1节
tennis shirt	网球衫	第1章1.3节
tension	张力	第4章2.3.1节
tension disc [dɪsk]	张力盘	第5章2.2.2节
tension roller	张紧辊	第5章2.1.4节
tex [teks]	特(号)数	第3章2.1节
texture ['tekstʃə]	膨体的,变形的	第3章1.1节
textured continuous multifilament	多孔变形丝	第3章2.1节
thigh	横裆	第2章3节
thigh [θaɪ] girth	大腿围长	第2章3节
thongs [θɒŋz]	(女)布带裤	第1章1.2节

thread breakage	断线	第6章3.2.2节
thread catcher	勾线器	第5章2.2.2节
thread end	线头	第5章1.1.10节
thread guard [gɑːd]	导线钩	第5章2.2.2节
thread hitching system	钩线系统	第5章2.2.2节
thread removal machine	清线头机	第5章2.3节
thread retainer [rɪ'teɪnə]	过线板	第5章2.2.2节
thread take-up system	挑线系统	第5章2.2.2节
thread tension	缝线张力	第6章3.2.2节
three-quarter sleeve	3/4长袖	第2章1.1.2节
three-thread overlock stitch	三线包缝	第4章2.3.1节
throat [θrəʊt] plate	针板	第5章2.2.2节
throughput ['θruːpʊt]	产量	第5章2.2.4节
ticket number	缝纫线标号	第3章2.1节
tie	领带	第1章1.1节
tie collar	领结领	第1章1.1节
tight fitting garment	紧身式服装	第4章2.4节
tights [taɪts]	紧身裤	第1章1.2节
tilt [tɪlt]	(使)倾斜	第5章1.1.5节
timer	定时器	第5章2.1.4节
tone in tone	配色	第3章2.1节
top	上装	第1章1节
top coat	(薄花呢)大衣	第1章1.1节
top collar ['kɒlə]	面领	第2章1.1.1节
top feed	上送料	第5章2.2.2节
top fly	面襟	第2章1.2.1节
top lapel width	驳头宽	第2章3节
top platen	上压板	第5章2.1.4节
top pressing	成衣熨烫	第5章1.1.11节
top sleeve	大袖	第2章1.1.1节
top stop	上止,头掣	第3章2.2.2节
topstitching ['tɒpstɪtʃɪŋ]	缉缝	第3章2.1节
toxic ['tɒksɪk]	有毒的	第3章2节
trace [treɪs]	描,描绘	第2章2节
tracing wheel	滚线轮	第5章2节
track [træk] suit	运动套装	第1章1.3节
trade policy	贸易政策	第1章2.3节
trademark ['treɪdmɑːk]	商标	第3章3.1.1节
transaction [træn'zækʃən]	交易	第4章1节
traverse ['trævə(ː)s]	(铺料设备)横移	第5章1.1.6节
trench [trentʃ] coat	军大衣式雨衣	第1章1.1节
Tributyltin [traɪ'bjuːtɪltɪn] (TBT)	三丁基锡	第6章3.1节
Tricot ['triːkəʊ] warp knitted fabric	特利考经编织物	第3章1.1节

trim	修整,修剪	第5章1.1.10节
trimmed edge	修剪过的布边	第4章2.3.1节
trimmings	辅料	第3章2节
trousers	长裤,裤子	第1章1.1节
trousers press	裤子压烫机	第5章2.3节
T-shirt	短袖针织圆领衫	第1章1节
tubular ['tjuːbjʊlə] form	圆筒状的	第5章1.1.6节
tuck stitched course	集圈横列	第1章1.1节
tucked dart	带褶子的省	第2章1.3节
tucks	(平行的)裥	第2章1.3节
tulip ['tjuːlɪp] sleeve	郁金花式短袖	第2章1.1.2节
tunnel ['tʌnl] finisher	隧道式整烫机	第5章2.3节
turn-ups	翻边	第1章1.2节
twill [twɪl]	斜纹	第3章1.1节
twist [twɪst]	加捻,捻度	第3章2.1节
two-piece sleeve	两片袖	第2章1.1.2节
two-thread overlock stitch	双线包缝	第4章2.3.1节
two-way pile	双向绒毛	第5章1.1.5节

u

under collar	底领	第2章1.1.1节
under fly	里襟	第2章1.2.1节
under pressing	中间烫,小烫	第5章1.1.10节
under sleeve	小袖	第2章1.1.1节
underarm ['ʌndərɑːm] dart	腋下的省	第2章1.3节
underclothing	内衣	第1章1节
underpants	(长)内裤	第1章1.2节
underwear ['ʌndəweə]	内衣	第1章1节
U-neck	U字领(圈)	第2章1.1.3节
unit production system	"单元"作业方法	第5章1.1.8节
United Nations Convention on Contracts for the International Sale of Goods (CISG)	联合国国际货物销售合同公约	第6章2节
unravel [ʌn'rævəl]	脱散,拆散,散开	第4章2.3.1节
upper arm (girth)	上臂围(围长)	第2章3节
upper front	上前片	第2章1.1.1节
upper garment	上装	第1章1节
upper hip (girth)	上臀围(围长)	第2章3节

v

V-bed machine	横机(V型针床机)	第3章1节
vacuum device	吸风装置	第5章2.1.1节
V-bed knitting machine	横机	第5章1.2节

vegetable fibre	植物纤维	第1章2.2节
Velcro ['vel'krəu]	尼龙搭扣	第3章2.2.4节
velvet ['velvɪt]	天鹅绒,丝绒	第5章1.1.5节
vent	开衩	第1章1.1节
vest	背心,内衣上装,西装马夹(美)	第1章1.1节
viscose ['vɪskəus]	黏胶纤维	第3章1.2节
Vislon zipper	维士隆(塑齿)拉链	第3章2.2.2节
V-neck	V字领(圈)	第2章1.1.3节
volume ['vɒljuːm] production	批量生产	第5章1.2节

w

w/o,(without)	没有,不计,不含	第2章3节
wadding ['wɒdɪŋ]	(纤维网状)填料	第3章2.4节
waist(girth)	腰围(围长)	第2章3节
waist dart	腰部的省	第2章1.3节
waist extended	拉开状态的腰围	第4章3节
waist relaxed	放松状态的腰围	第4章3节
waistband	腰头	第1章1.1节
waistband depth	腰头宽	第2章3节
waistcoat ['weɪstkəut]	西装马夹	第1章1.1节
wale [weɪl]	(针织物)纵行	第2章2节
wale density	纵行密度(横密)	第5章1.2节
warning ['wɔːnɪŋ] label	警示标签	第3章3.1.5节
warp [wɔːp]	经纱	第2章2节
warp knitted meshes	经编网眼布	第3章1.2节
washing label	洗涤标签,(洗水唛)	第3章3.1节
watch pocket	表袋	第2章1.1.4节
waxing ['wæksɪŋ]	上蜡	第5章1.2节
wedge [wedʒ] dart	契型省	第2章1.3节
weft knitted pile fabric	纬编割绒织物	第3章1.2节
welding ['weldɪŋ]	焊接	第2章1节
welt	袋贴边	第2章1.1.4节
welt [welt] pocket	贴边袋	第2章1.1.4节
whisker ['hwɪskə] hand brushing	须状手擦(俗称"猫须")	第1章1.2节
widening or narrowing	放针或收针	第1章2.3节
width of wrap overlay ['əuvəleɪ]	上叠片宽	第2章3节
width of wrap underlay ['ʌndəleɪ]	下叠片宽	第2章3节
width variation [ˌveərɪ'eɪʃən]	门幅变化	第5章1.1.4节
wind-cheater	防风上衣	第1章1.1节
winding ['waɪndɪŋ]	络纱	第5章1.2节
women's coat	女装风衣	第1章1.3节
wool	毛	第1章2.2节
woolen Melton ['meltən]	粗纺毛麦尔登呢	第3章1.1节

working capacity [kə'pæsıtı]　　生产能力　　　　　　第5章1.1.5节
working drawing　　　　　　　　效果图　　　　　　　第4章2.2节
working dress　　　　　　　　　工作服　　　　　　　第1章1.1节
working needle　　　　　　　　　参与工作的织针　　　第5章1.2节
working sheet　　　　　　　　　工艺单　　　　　　　第4章1节
working sketch　　　　　　　　　效果图　　　　　　　第4章2节
working specification sheet
(SPEC sheets)　　　　　　　　　工艺单　　　　　　　第1章3.2节
workmanship ['wɜːkmənʃıp]　　做工　　　　　　　　第5章1.1.12节
workstation　　　　　　　　　　工位　　　　　　　　第5章1.1.8节
World Trade Organization(WTO)　世界贸易组织　　　　第1章2.3节
worn [wɔːn] clothing　　　　　　旧衣物　　　　　　　第1章2.1节
woven ['wəʊvən] fabric ['fæbrɪk]　机织织物　　　　　　第1章1.1节
wrapped-around [ræpt- ə'raʊnd]
buttons　　　　　　　　　　　　包扣　　　　　　　　第3章2.2.1节
wrap-round skirt　　　　　　　　左右裙前片叠合的裙子，
　　　　　　　　　　　　　　　　围裹裙　　　　　　　第1章1.2节
wrinkle　　　　　　　　　　　　褶皱　　　　　　　　第6章3.2.2节
wrong side (WS)　　　　　　　　织物反面　　　　　　第2章1.1.1节

x

x-y coordinate [kəʊ'ɔːdɪnɪt]　　x-y 坐标　　　　　　第5章1.1.3节

y

yacht [jɒt] club　　　　　　　　游艇俱乐部　　　　　第1章1.1节
yarn count　　　　　　　　　　纱支　　　　　　　　第4章2.1节
yarn dyed　　　　　　　　　　　色织　　　　　　　　第3章1.1节
yarn package　　　　　　　　　纱线的卷装　　　　　第5章1.2节
Y-fronts　　　　　　　　　　　（男）(前开裆)短内裤　第1章1.2节
yoke [jəʊk]　　　　　　　　　　覆肩　　　　　　　　第1章1.1节
　　　　　　　　　　　　　　　裙(或裤)的拼腰　　　第1章1.2节

z

zip [zɪp] / zipper ['zɪpə]　　　　拉链　　　　　　　　第2章1.2.1节
zip slider　　　　　　　　　　　拉链头　　　　　　　第3章2.2.2节
zipped closure ['kləʊʒə]　　　　用拉链的闭口　　　　第1章1.2节
zipped-on　　　　　　　　　　用拉链脱卸的　　　　第2章1.2.1节
zipper guard [gɑːd]　　　　　　拉链挡布　　　　　　第3章2.2.2节
zipper tape　　　　　　　　　　拉链(底)带　　　　　第3章2.2.2节
zipper teeth　　　　　　　　　拉链齿　　　　　　　第3章2.2.2节